Undergraduate Lecture Notes in Physics

For further volumes:
http://www.springer.com/series/8917

Undergraduate Lecture Notes in Physics (ULNP) publishes authoritative texts covering topics throughout pure and applied physics. Each title in the series is suitable as a basis for undergraduate instruction, typically containing practice problems, worked examples, chapter summaries, and suggestions for further reading.

ULNP titles must provide at least one of the following:

- An exceptionally clear and concise treatment of a standard undergraduate subject.
- A solid undergraduate-level introduction to a graduate, advanced, or non-standard subject.
- A novel perspective or an unusual approach to teaching a subject.

ULNP especially encourages new, original, and idiosyncratic approaches to physics teaching at the undergraduate level.

The purpose of ULNP is to provide intriguing, absorbing books that will continue to be the reader's preferred reference throughout their academic career.

Series Editors

Neil Ashby
Professor, Professor Emeritus, University of Colorado Boulder, CO, USA

Professor, William Brantley
Furman University, Greenville, SC, USA

Michael Fowler
Professor, University of Virginia, Charlottesville, VA, USA

Michael Inglis
Professor, SUNY Suffolk County Community College, Selden, NY, USA

Elena Sassi
Professor, University of Naples Federico II, Naples, Italy

Helmy Sherif
Professor, University of Alberta, Edmonton, AB, Canada

William M. Hartmann

Principles of Musical Acoustics

 Springer

William M. Hartmann
Michigan State University
East Lansing, MI, USA

ISSN 2192-4791   ISSN 2192-4805 (electronic)
ISBN 978-1-4614-6785-4   ISBN 978-1-4614-6786-1 (eBook)
DOI 10.1007/978-1-4614-6786-1
Springer New York Heidelberg Dordrecht London

Library of Congress Control Number: 2013938260

Printed on acid-free paper

Springer is part of Springer Science+Business Media (www.springer.com)

To Mitra and Daniel

Preface

Musical acoustics is a scientific discipline that attempts to put the entire range of human musical activity under the microscope of science. Because science seeks understanding, the goal of musical acoustics is nothing less than to understand how music "works," physically and psychologically. Accordingly, musical acoustics is multidisciplinary. At a minimum it requires input from physics, physiology, psychology, and several engineering technologies involved in the creation and reproduction of musical sound.

As a scientific discipline, musical acoustics poses several questions. The first question has to be, "Why would anyone want to study it?" Unlike some other scientific pursuits, it is unlikely to improve the health of the population, nor is it likely to be the engine that drives the economy to new heights. The answer to the question lies within the appeal of music itself. Music is fundamentally important to human beings—it has been a part of every culture throughout history. Nothing has greater capability to inspire human emotions. Thus musical acoustics is an attempt to take a *rational* approach to a human activity that is essentially emotional. In the end, music is compelling. It compels us to listen to it, to create it, and ultimately to try to understand it.

Musical acoustics is an ancient science—the rational approach to music was part of the ancient Greek quadrivium. Musical acoustics also claims an ancient technology. For instance, the greatest violins in history were made in the eighteenth century. Is there any other scientific discipline where the peak of a relevant technology occurred more than 200 years ago? At the same time, musical acoustics is modern. New electronic methods of producing, processing, and recording music appear with every new year. The discipline forms an important part of our rapidly developing worldwide communications technology, and the study of music processing by the human brain is an important aspect of contemporary neuroscience.

The teaching and learning of musical acoustics likewise presents some challenges. As an ancient science, with millions of participants as performers and instrument makers, and a technology that is as much art as it is science, the subject abounds in details. Every musical idea and every musical instrument have seen dozens of variations—some enduring, others fleeting curiosities. These details are often fascinating, and they are part of the charm of the subject. Musicians are justifiably proud of their rich musical culture. On the other hand, students approaching the subject for the first time can easily be overwhelmed by all the details. As its name implies, this little book, *Principles of Musical Acoustics*, focuses on the basic principles in the science and technology of music. Musical examples and specific musical instruments demonstrate the principles.

This book begins with a study of vibrations and waves—in that order. These topics constitute the basic physical properties of sound, one of two pillars supporting the science of musical acoustics. The second pillar is the human element, the physiological and psychological aspects of acoustical science. The perceptual topics include loudness, pitch, tone color, and localization of sound.

With these two pillars in place, it is possible to go in a variety of directions and in no particular order. The book treats in turn the topics of room acoustics, audio—both analog and digital—broadcasting, and speech. It ends with chapters on the traditional musical instruments, organized by family.

The mathematical level of this book assumes that the reader is familiar with elementary algebra. Trigonometric functions, logarithms, and powers also appear in the book, but computational techniques are included as these concepts are introduced, and there is further technical help in appendices.

There are exercises at the end of each chapter. Most of the exercises follow in a straightforward way from material presented in the chapter. Other exercises are a bit of a stretch. Both kinds are useful. Rather complete answers to most of the exercises appear in the back of the book, but anyone who wants to use this book for serious study ought to try to do the exercises before looking up the answers in the back. The goal, of course, is active involvement with the material in the text. Working an exercise, even if unsuccessfully, is of more value than reading a text passively.

The elements in this book have been developed over the course of several decades in teaching musical acoustics in the Department of Physics and Astronomy at Michigan State University. I am grateful to the thousands of students who have experienced the course and given me inspiration, advice, data, cheers, complaints, and other forms of feedback. Especially, I am particularly grateful to Professor Jon Pumplin, Dr. Diana Ma, and Mr. John McIntyre for helpful suggestions. Tim McCaskey, Nick Nuar, Ben Frey, Louis McLane, and Yun Jin Cho helped me assemble notes into this volume. I could not have written this book without their help. Finally, I am grateful to my wife, Christine Hartmann, for her support over the years.

East Lansing, MI, USA   William M. Hartmann

♠

Contents

♠

Chapter 1
Sound, Music, and Science

Sound is all around us. It warns us of danger, enables us to communicate with others, annoys us with its noise, entertains us by radio and iPods, and captivates us in music. It is an important part of the daily lives of hearing people.

We are concerned here with a *science* of sound, especially musical sound. There is something of a paradox in this concept. Sound itself tends to be personal and it often brings up an emotional response. On the other hand there is science, supposedly rational and objective. Merging these two aspects of human experience is the science and art of acoustics.

Acoustics is foremost a science. As such it is quantitative, attempting to account for the physical world and our perception of it in ways that can be measured with experiments and described with mathematical models. A quantitative science like acoustics operates with a number of ground rules:

- *Definitions*: We need precise definitions for ideas and for quantities. We often take common words and give them meanings that are more tightly constrained than in everyday speech. For instance, in the next chapter the word "period" will be given a precise mathematical meaning.
- *Simplification*: The real world is complicated. Science gains its power by simplification. For instance, everyday materials are complex compounds and mixtures of dozens of chemical elements, but the chemist uses pure chemicals to gain control of his experiment. In the same way, speech and music are complicated signals, but the acoustician uses signals that are no more complicated than necessary for the intended purpose.
- *Idealization*: The technique of idealization is like simplification in that it is a scientific response to a messy world. Idealization applies to the conceptual models that we use to explain some aspect of the world. An idealized model attempts to capture the essence of something, even though the model may not explain every detail. For instance, there is the concept that planets, like the Earth and Mars, orbit the Sun because of the Sun's strong gravitational attraction. That is a powerful model of our solar system. But, it is not a perfect model because the

W.M. Hartmann, *Principles of Musical Acoustics*, Undergraduate Lecture Notes
in Physics, DOI 10.1007/978-1-4614-6786-1_1,
© Springer Science+Business Media New York 2013

planets attract one another too, and this is not included in the basic heliocentric model of orbital motion. Nevertheless, the model successfully abstracts the most important character of planetary motion from a complicated real-world situation. This model is a useful idealization.

And so we begin with the science of acoustics ...

An acoustical event consists of three stages as shown in Fig. 1.1: First, the sound is generated by a *source*. Second, the sound is *transmitted* through a medium. Third, the sound is intercepted and processed by a *receiver*. These three stages form the basis of acoustical science. We consider them in turn and **highlight** items that we expect to study in detail.

1.1 The Source

The source of a sound is always a vibration of some kind. For example, it might be the vibration of a drumhead. A drum is a **traditional musical instrument**, and our approach makes a study of traditional musical instruments by their families: brass instruments, woodwind instruments, string instruments, and percussion instruments.

The **human voice** is another source of sound, arguably the most important of all. The basic science of the human voice is enormous fun because a few simple principles serve to take us a long way toward understanding what is going on acoustically.

A more modern source of sound is the **loudspeaker**. It too is a vibrating system, but unlike the other vibrating systems, it is not caused to vibrate directly by human action. The loudspeaker is a **transducer** that converts electronic signals into acoustical signals. This marriage of electronics and acoustics is called **audio**; its technological and cultural significance is so important that it is impossible to imagine modern life without it. It is closely allied with **broadcasting** by radio and TV. Audio technology has made possible two other kinds of musical instrument. One is the electrified instrument, where the vibrations of a physical object are converted into electronic form. For instance, the vibrations of a guitar string can be converted by a pickup into an electrical signal. The other is an **electronic instrument**, analog or digital, where the original vibrations are generated electronically.

Vibrations are so basic in the study of acoustics that the first chapters are dedicated to developing the terminology and basic mathematical relations for the study of vibration. These chapters also introduce electronic **instrumentation** used to study vibrations.

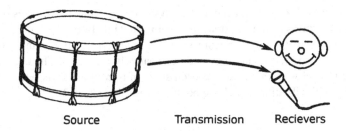

Source　　　　Transmission　　　Recievers

Fig. 1.1 A sound from a source is transmitted to two receivers, a human listener and a microphone

1.2 Transmission

The vibration of the drumhead causes the air around it to vibrate. This vibration propagates as a wave through the air. Accordingly, the study of waves occupies an important place in musical acoustics. The physics of waves is actually a long and deep subject because there is a rich variety of **wave phenomena**. We shall deal with some of the most fundamental properties. Wave motion is not only a characteristic of sound waves (acoustics), but also characterizes the transmission of light and radio waves. The wave principles that one learns in studying acoustics apply directly to optics (light) and electromagnetic radiation in the form of radio waves. Therefore, by learning about the weird things that can happen to an acoustical wave you immediately understand something about optical mirages and problems with your cell phone.

The transmission of sound from a source to a receiver does not take place entirely by a straight line path. The sound waves are reflected from the walls of a room and by other surfaces in the room. The character of the room puts its indelible stamp on the sound wave as it is finally received. This is the subject of **room acoustics**. It covers a lot of ground, from the problem of noise in your residence to the design of multi-million-dollar concert halls.

1.3 Receiver

The most important receiver of sound is the human ear and brain—the human auditory system. Sound waves, of the kind that we study in the musical acoustics, are meant to be heard, understood, and appreciated. In the final analysis, the strengths and limitations of the human auditory system determine everything else we do in acoustics. There are two basic divisions of subject matter in the study of human hearing; the first is the **anatomy** and **physiology** of the auditory system, the second is the **psychoacoustics**. The anatomy and physiology describe the tools we have to work with as listeners; the psychoacoustics describes the function of these tools, converting sound waves into perceptions. Important perceptual properties of sound include the **loudness** of tones, **pitch, tone color**, and **location**.

These three stages: source, transmission, and reception, appear in any acoustical experiment or experience, and they can be separately identified. The chapters that follow try to deal with the details of each stage in turn. It all starts with a source, specifically with vibration, which we begin in earnest in Chap. 2.

Comparisons

We make comparisons everyday. It is part of living. Some things are better—other things are worse. In a science like acoustics comparisons are usually quantitative. This section describes the most important quantitative comparisons.

Differences:

Quantitatively, difference means subtraction. We obtain a difference by subtracting two values. Thus if A has a length of 2 m and B has a length of 2.1 m, then the difference between B and A is a length of $2.1 - 2 = 0.1$ m.

You will notice that a difference has units—such as meters. It has the units of the quantities that are subtracted. It follows that we can only take differences between quantities that have the same kind of units. For instance, it is not possible to find the difference between 2 m and 3 kg.

It is possible, however, to find the difference between 2 m and 210 cm. Although meters and centimeters are different units, they are the same *kind* of units. They are both lengths. Still, one cannot take the difference directly. The difference calculation $210 - 2$ would give a nonsensical answer. To take differences of quantities expressed in different units of the same kind requires a conversion of units so that the quantities being subtracted have identical units. In this instance we might chose to convert 210 cm to 2.1 m and take the difference $2.1 - 2.0 - 0.1$ m, as before. Alternatively, we might choose to convert 2 m to 200 cm and take the difference $210 - 200 = 10$ cm. Either way is correct.

Ratios

Quantitatively, ratio means division. Two quantities are compared by dividing one by the other. Thus if A has a length of 2 m and B has a length of 2.1 m, then the ratio of B to A is $2.1/2.0 = 1.05$. Another word that is often used in connection with the ratio concept is the word "factor." We say that B is greater than A by a factor of 1.05. The implication of this statement is that we can find out how big B is by starting with A and multiplying by a factor of 1.05. If "ratio" means divide, then "factor" means multiply.

You will notice that a ratio does not have any units. That makes it very different from a difference. So whereas $2.1 - 2.0 = 0.1$ and $210 - 200 = 10$, the ratio $2.1/2.0$ equals 1.05 and the ratio $210/200$ also equals 1.05. We can go further: A length of 2.1 m is 82.677 in. and a length of 2.0 m is 78.74 in. The ratio $82.677/78.74$ is also equal to 1.05. Coming out in a unitless way like this gives the ratio comparison a certain conceptual advantage over the difference comparison. But in order for the ratio comparison truly to have no units, the quantity in the numerator and the quantity in the denominator must have the *same* units. Thus it would be wrong to find the ratio of $(210\,cm)/(2\,m) = 105$. The number 105 is not a correct ratio.

Of course, there are division operations that are not ratio comparisons. For example 100 km driven in 2 h corresponds to an average speed of 50 km/h. Dividing 100 km by 2 h leads to a speed which has physical units and therefore is not a ratio comparison.

Percentages

Percentages are just like ratios. If A is 2.0 and B is 2.1, then the ratio of A to B is 2.0/2.1, which is about 0.952. We say that A is 95.2 % of B. Alternatively we could calculate the ratio $2.1/2.0 = 1.05$ and say that B is 105 % of A. Either way, we get a percentage comparison by multiplying a ratio by 100.

Percentage Change

The percentage *change* is a more common comparison than the percentage comparison. The percentage change combines the concepts of difference and ratio. It is the difference between two quantities divided by one of those two original quantities. In terms of a fraction, the difference is in the numerator and one of the two quantities is in the denominator. The quantity in the denominator is the reference quantity. The concept of change is that we start with something and end up with something else. The reference quantity is what we start with.

In terms of quantities A and B, the percentage change from A to B has A in the denominator.

$$Percent\ change = (B - A)/A \tag{1.1}$$

We can apply this equation to a change in height.

Example 1: At age 16 Shaquille O'Neal was $A = 201\,cm$ tall. By age 21 he had grown to be $B = 216\,cm$ tall. The percentage change is a growth of $(216 - 201)/201 = 0.075$ or 7.5 %. The reference quantity is the height in the starting year, namely 201 cm.

Example 2: Alternatively we might start with a sandwich that is $A = 2.1\,m$ long and nibble on it until it is $B = 2.0\,m$ long. The percentage is a decrease, namely a negative change, $(2.0 - 2.1)/2.1 = -0.0476$, a decrease of 4.76 %. The reference quantity is the starting sandwich length of 2.1 m.

There is a simple relationship between the ratio and the percentage change indicated by the end of Eq. (1.2)

$$Percent\ change = (B - A)/A = B/A - 1. \tag{1.2}$$

Equation (1.2) shows that the percentage change is always the difference between a ratio (like B/A) and the number 1. That little bit of algebra will prove helpful in some of the exercises later in this book. Like a ratio, the percentage change has no units.

Example 3: At the start of the year 2009, the Dow Jones Industrial Average of stock prices was 8776. At the end of the year, it was 10428. The change for 2009 was

$$10428./8776. - 1. = 0.188 \tag{1.3}$$

or 18.8 %.

Powers of ten

Powers of ten are illustrated with their prefixes by introducing the fictitious unit of the Snurk.

One gigaSnurk $= 10^9$ Snurks $= 1{,}000{,}000{,}000$ Snurks (one billion)*
One megaSnurk $= 10^6$ Snurks $= 1{,}000{,}000$ Snurks (one million)
One kiloSnurk $= 10^3$ Snurks $= 1{,}000$ Snurks (one thousand)
One milliSnurk $= 10^{-3}$ Snurks $= 1/1000$ Snurks (one thousandth)
One microSnurk $= 10^{-6}$ Snurks $= 1/1{,}000{,}000$ Snurks (one millionth)
One nanoSnurk $= 10^{-9}$ Snurks $= 1/1{,}000{,}000{,}000$ Snurks (one billionth)

Scientific notation

$21{,}000$ Snurks $= 2.1 \times 10^4$ Snurks
0.00345 Snurks $= 3.45 \times 10^{-3}$ Snurks

*As of 1975, 1,000,000,000 is both the American and the British billion.

Exercises

Exercise 1, Classification
   In the categories of Source, Transmission, and Receiver, how would you classify a reflecting wall? A musical instrument? A microphone?

Exercise 2, Sound goes everywhere
   Sound not only propagates in air, but also propagates in water and even in solids. What good is that?

Exercise 3, Bad conditions
   Under what conditions have you had difficulty hearing something that you wanted to hear?

Exercise 4, Idealized models

Sam says, "Models of the world are essential for human thought. Without models every experience would be a brand new event. Models provide a context in which experiences fit, or don't fit."

Pam says, "Models of the world can be dangerous. They lead to assumptions about events and people that may not be true. Models can prevent you from seeing things as they truly are."

Defend Sam or Pam, or both.

Exercise 5, More waves

The text mentions sound waves, light waves, and radio waves. What other kinds of waves do you know?

Exercise 6, More Shaq

At age 16 Shaquille O'Neal weighed 120 kg. At age 21 his weight was 137 kg. Calculate the difference, the ratio, and the percentage of change over those 5 years.

Exercise 7, Investment experience

On Monday your investment loses 10 % of its value. However, on Tuesday your investment gains 10 %. Do those two changes perfectly cancel each other? (a) Show that these two changes actually cause you to lose 1 %. (b) Would the net result be different if you gained 10 % on Monday and lost 10 % on Tuesday?

Exercise 8, Powers of ten

Recall that $10^3 \times 10^5 = 10^8$. Now evaluate the following:
(a) $10^3 \times 10^2 = ?$ (b) $10^3/10^2 = ?$ (c) $10^3 \times 10^{-2} = ?$ (d) $10^3/10^{-2} = ?$ (e) $10^3 + 10^2 = ?$

Exercise 9, Scientific notation

Express the following in scientific notation: (a) $20 billion. (b) Ten cents. (c) 231,000. (d) 0.00034.

Exercise 10, micro, milli, kilo, mega

(a) How many seconds in a megasecond? (b) How many microseconds in a second? (c) How many milliseconds in a kilosecond?

♠

Chapter 2
Vibrations 1

Sound begins with vibrations. For example, the sound of a guitar begins with the vibration of the strings, and the sound of a trumpet begins with the vibrations of the player's lips. Perhaps less obvious, the sound of the human voice begins with the vibration of vocal folds (vocal cords), and sounds that are reproduced electronically are ultimately converted into acoustical waves by the vibration of a loudspeaker cone. Because sound starts with vibrations, it's natural that the study of acoustics begins with the study of vibrations.

Fundamentally, a vibration is a motion that is back and forth. Vibrations can be simple, like the swinging of a pendulum, or they can be complicated, like the rattle of your car's glove box when you drive on a bumpy road. Here we follow a path that is traditional in science; we consider first the simplest possible system and try to describe it as completely as possible. This vibration is called *simple harmonic motion*.

2.1 Mass and Spring

2.1.1 Definitions

The most fundamental mechanical vibrator is a single mass hanging from a spring. If there is no motion (no vibration) the force of gravity pulling the mass down is just balanced by the force of the spring pulling up. The mass is at rest. It is said to be at *equilibrium*. If the mass moves away from its equilibrium position it has a *displacement*. If the mass moves up above the equilibrium position, the displacement is positive, if it moves down, the displacement is negative. When the mass is at equilibrium the displacement is zero (Fig. 2.1).

If the spring and mass system is allowed to vibrate freely, the mass will undergo simple harmonic motion. As the mass vibrates up and down, its displacement is

W.M. Hartmann, *Principles of Musical Acoustics*, Undergraduate Lecture Notes
in Physics, DOI 10.1007/978-1-4614-6786-1_2,
© Springer Science+Business Media New York 2013

Fig. 2.1 The spring and
mass system. In equilibrium
the displacement is zero.
When the mass moves above
the equilibrium point the
displacement is positive.
When the mass moves below
the equilibrium point the
displacement is negative

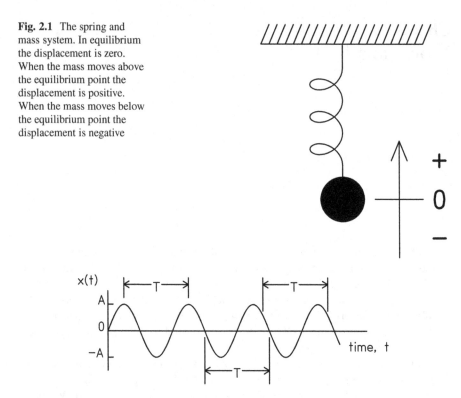

Fig. 2.2 The sine wave has extreme values $\pm A$, called the amplitude. It has a characteristic waveform shape as shown here. The sine wave is periodic, so that any two equivalent points on the waveform are separated in time by the time interval T known as the "period." Theoretically, the sine wave extends indefinitely to infinite positive and infinite negative time, having no beginning or end

alternately positive and negative. The displacement (x) can be plotted as a function of time (t) on a graph like Fig. 2.2.

The maximum displacement is called the *amplitude*. The most negative displacement is the same number with a minus sign in front of it. Simple harmonic motion is an example of *periodic* motion. That means that the motion is repeated indefinitely. There is a basic pattern, called a *cycle*, that recurs. The duration of the cycle is called the *period*. It is a time and it is measured in seconds, or milliseconds (1 one-thousandth of a second).

2.1.2 How It Works

To start the vibrations of a mass and spring system, you displace the mass from its equilibrium position and let it go. Let us suppose that you gave the mass a negative (downwards) displacement, stretching the spring. Because the spring is stretched, there is a *force* on the mass trying to bring the mass back to its equilibrium position. Therefore, the mass moves toward equilibrium. Soon the mass arrives at the equilibrium position, and at that point there is no more force. You might think that the mass would just stop at equilibrium, but this is not what happens. The mass doesn't stop because in the process of coming back to equilibrium, the mass has acquired some speed and, precisely because it is massive, it has acquired some *momentum*. Because of the momentum, the mass overshoots the equilibrium point, and now it has a positive displacement. The positive displacement causes the spring to be compressed, which ultimately forces the mass to move back down again toward equilibrium. As the mass passes the equilibrium point for a second time, it now has momentum in the opposite direction. That momentum keeps the mass going down until it has reached the negative amplitude that it had at the very start of this story. The mass and spring system has now executed one complete cycle. The amount of time needed to complete the cycle is one period.

The workings of a simple harmonic vibrating system like the mass and spring illustrate some very general principles. First, it seems clear that if we have a very stiff spring, then any displacement from equilibrium will lead to a large force and that will accelerate the mass back toward equilibrium faster than a weak spring. Therefore, we expect that making the spring stiffer ought to make the motion faster and decrease the period. Second, it also seems clear that if the mass is very massive, then it will respond only sluggishly to the force from the spring. (Did you know that the English-system unit for mass is the *slug*? How appropriate!) Therefore, we expect that making the mass heavier ought to make the motion slower and increase the period. As we will see in Chap. 3, that is exactly what happens.

2.2 Mathematical Representation

Mathematically, simple harmonic motion is described by the sine function of trigonometry. Therefore, simple harmonic motion is sometime called "sine wave motion," and the waveform shown in Fig. 2.2 is called a *sine* wave. It is described by the equation

$$x(t) = A \sin(360\,t/T + \phi) \tag{2.1}$$

where A is the amplitude, T is the period in seconds, and ϕ is the *phase* in degrees. [Note: Symbol ϕ is the Greek letter *phi*. The names of all Greek letters are in Appendix H.] A review of trigonometry, particularly the sine function, is given in Appendix B.

We will spend some words dissecting equation (2.1). The units of the amplitude A are the same as the units of x, in our case, mechanical displacement. It might be measured in millimeters. If the sine wave $x(t)$ describes some other physical property, such as air pressure or electrical voltage, the units of the amplitude change accordingly. By convention, A is always a positive real number. Amplitude A can be zero, but in that case there is no wave at all, and there is nothing more to talk about. The amplitude multiplies the sine function—**sin**—which has a maximum value of 1 and a minimum value of -1. It follows that the sine wave $x(t)$ has a maximum value of A and a minimum value of $-A$.

The sine function is a function of an angle, i.e., whenever you see an expression like $\sin(\ldots)$, the quantity..., which is the *argument* of the sine function, has to be an angle. Therefore, "$360t/T + \phi$" is an angle. The special phase ϕ in the equation is the value of the angle when $t = 0$. In some cases we are interested in the function only for positive values of time t, it being presumed that the wave starts at time $t = 0$. Then phase ϕ is called the "starting phase."

The sine function is periodic; it repeats itself when the angle (or argument) changes by $360°$. Equation (2.1) shows that if the running time variable t starts at some value and then increases by T seconds, the function comes back to its starting point. That is why we call T the period.

The reciprocal of the period leads us to the definition of the frequency (f) of the sine wave,

$$f = 1/T, \tag{2.2}$$

and its units are cycles per second, or Hertz, (abbreviated Hz). Therefore, one can write the sine wave in another form,

$$x(t) = A \sin(360\, ft + \phi). \tag{2.3}$$

Example: As an example, we consider the sine-wave vibration of an object whose position in millimeters is given by

$$x(t) = 4 \sin(360 \cdot 60\, t + 90). \tag{2.4}$$

From Eq. (2.1), the amplitude of the vibration is $4\,mm$, which means that the maximum positive and negative excursions are $\pm 4\,mm$. The frequency is $60\,Hz$, and the starting phase is $90°$. See Fig. 2.3.

The argument of the sine function, here given by $360ft + \phi$, is an angle called the *instantaneous phase*. We will give it the symbol Θ. It is assumed that angle Θ is always expressed in units of degrees. In a complete circle there are $360°$.

The role of the sine function as a periodic function of a time-dependent instantaneous phase is emphasized by separating the aspects of periodicity and time dependence. One simply writes the sine wave in the form

$$x(t) = A \sin(\Theta) \tag{2.5}$$

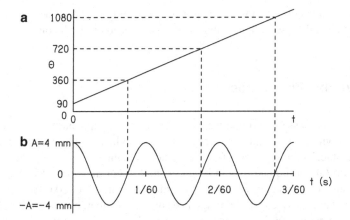

Fig. 2.3 Panel (**a**) shows the instantaneous phase angle Θ, increasing from an initial value $\phi = 90$ as time goes on. Over the duration shown, this angle advances through three multiples of 360°. Panel (**b**) shows what happens when one takes the sine of angle Θ and multiplies by the amplitude A. Here the amplitude A was chosen to be 4 mm

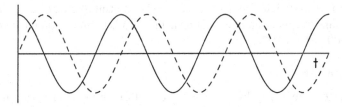

Fig. 2.4 The wave shown by the *solid line* is said to lead the wave shown by the *dashed line* because every waveform feature—peak, positive-going zero crossing, etc.—occurs at an earlier time for the solid line. Alternatively the dashed line wave can be said to lag the solid line wave. Both waves have the same frequency and amplitude, but their starting phases are different

where Θ is the instantaneous phase angle, measured in degrees, as shown in Appendix B. Here, angle Θ is a function of time

$$\Theta = \Theta(t) = 360ft + \phi \qquad (2.6)$$

As t increases, the angle Θ increases linearly, and the sine function goes through its periodic oscillations, as shown in Fig. 2.3. Figure 2.3 shows the special case where the phase is $\phi = 90°$.

Phase Lead–Phase Lag When two sine waves have the same frequency but different starting phases, one of them is said to lead or to lag the other. Figure 2.4 is an example. The lagging wave is given by $x(t) = A\sin(360ft)$ and the leading wave is given by $x(t) = A\sin(360ft + 90)$. According to Eqs. (2.5) and (2.6) the starting phase angle for the leading wave is $\phi = 90°$. This angle is positive and less than 180°, which corresponds to a condition for leading. It should be evident that if ϕ had been equal to 180°, then neither wave would lead or lag the other.

2.3 Audible Frequencies

The spring and mass system that we used to introduce simple harmonic motion lets you see the vibration but the frequency is too low to hear. Such vibration is said to be "infrasonic." Vibrations of a few Hertz can sometimes be felt, and they can do damage too (an earthquake would be an example), but they cannot be heard. The audible range of frequencies is normally said to be from 20 to 20,000 Hz, or 20 Hz to 20 kHz. That statement of the range is easy to remember, but the practical range tends to be smaller. Without a special acoustical system it is not possible to hear 20 Hz. A frequency of 30 Hz is a more practical lower limit. Many college students cannot hear 20,000 Hz either. A more realistic upper limit of hearing is 17,000 Hz. We are quite accustomed to doing without the lowest and highest frequencies. Many loudspeakers that claim to be high quality cannot reproduce sounds below 100 Hz. FM radio stations are not even allowed to transmit frequencies higher than 15,000 Hz, and telephone communication in the USA is normally limited to the band between 300 and 3,300 Hz.

Octave Measure

To change a frequency by an octave means to multiply it or divide it by the number 2. Starting with a frequency of 440 Hz, going up one octave gets you to 880 Hz and going up another octave gets you to 1,760 Hz. The sequence is continued in Fig. 2.5. Again starting at 440 Hz, going down an octave leads to 220 Hz and going down another octave leads to 110 Hz. The measure of an octave is a basic element in the music of all cultures.

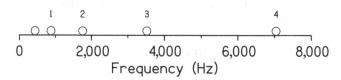

Fig. 2.5 The horizontal axis is a linear frequency scale. The *circles* show 440 Hz and octaves above 440, namely 880, 1,760, 3,520, and 7,040 Hz. With increasing octave number the frequencies become more widely spaced on the linear scale

Exercises

Exercise 1, Human limits

Nominally, the limits of human hearing are 20 and 20,000 Hz. Find the periods of those two waves.

Exercise 2, A low frequency

The second hand of a clock takes 60 s for one cycle. What is the frequency in Hertz?

Exercise 3, Time conversions

A millisecond is 1 one-thousandth of a second, and a microsecond is 1 one-millionth of a second. (a) How many milliseconds is 2 s? (b) How many milliseconds is 30 μs?

Exercise 4, Frequency conversions

A kilohertz (kHz) is 1,000 Hz. (a) How many kilohertz is 16,384 Hz? (b) How many hertz is 10 kHz?

Exercise 5, Period and frequency

(a) If the period is 1 ms, what is the frequency in kHz? (b and c) If the frequency is 10,000 Hz, what is the period in milliseconds and in microseconds?

Exercise 6, Conditions for phase lag

The mathematical conditions for phase leading are described in the text. From this, can you infer the conditions for lagging?

Exercise 7, Optical analogy

If an "infrasonic" sound has a frequency that is too low to hear, what is "infrared" light?

Exercise 8, Telephone bandwidth

The telephone bandwidth is from 300 to 3,300 Hz. How many octaves is that bandwidth?

Exercise 9, Sine waves

The sine wave in Fig. 2.2 has a starting phase of zero. It is reproduced in Fig. 2.6. (a) On the same set of axes, draw a sine wave with the same frequency and same amplitude but with a starting phase of 180°. (b) Draw a sine wave with the same amplitude but twice the frequency. (c) Draw a sine wave with the same frequency but half the amplitude. (d) The sine wave is said to be a "single valued function," meaning that for every point in time there is one and only one value of the wave. Sketch a function that is not single valued for comparison.

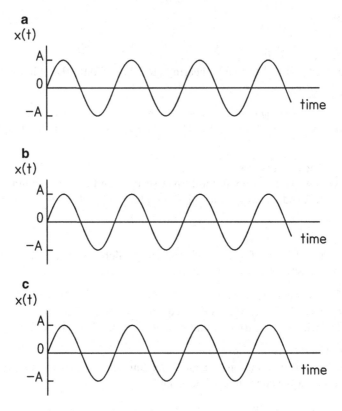

Fig. 2.6 Three practice waves for Exercise 9

Chapter 3
Vibrations 2

Chapter 2 introduced the concept of simple harmonic motion. This concept allowed us to define the terms used to discuss vibrating systems—terms like amplitude and frequency. Chapter 3 extends this discussion to additional properties of real vibrating systems. It ends with the topics of spectrum and resonance, essential concepts in any form of physics, especially acoustics.

3.1 Damping

Ideal simple harmonic motion goes on forever. It is a kind of perpetual motion machine with an amplitude that never changes. It just keeps on going and going and going. Real, passive mechanical systems, like a mass suspended from a spring, are not like this. The vibrations (or oscillations) of a real, free mechanical system are damped by frictional forces, including air resistance. The amplitude of such systems is not constant but decreases gradually with time. Figure 3.1 shows motion that would be simple harmonic (sine wave) motion except that it is damped.

Because of damping in real systems, our theoretical picture of simple harmonic motion is an idealization. Idealizations are common in science. They help us deal with certain truths about the world without the encumbrances of a lot of messy real-world details. The ideal simple harmonic motion is a useful idealization because many systems vibrate for many cycles before decaying appreciably. Also, it is often possible to compensate for the damping (or decay) of vibrations by adding energy to the system from outside. Then the system is *active*, not passive or freely vibrating.

W.M. Hartmann, *Principles of Musical Acoustics*, Undergraduate Lecture Notes in Physics, DOI 10.1007/978-1-4614-6786-1_3,
© Springer Science+Business Media New York 2013

Fig. 3.1 Rapidly damped simple harmonic motion—like clinking glasses

3.2 Natural Modes of Vibration

The concept of "modes of vibration" is important in mechanics and acoustics. You already know something about it. The spring and mass system, with the mass confined to move in one dimension, is a system that has one mode of vibration. A mode is described by its properties:

1. *Frequency*: Most important, a mode has a specific frequency. The spring and mass system has a frequency given by the following formula:

$$f = \frac{1}{2\pi} \sqrt{\frac{s}{m}} \tag{3.1}$$

Here s is the stiffness of the spring, measured in units of Newtons per meter. A Newton is the unit of force in the metric system of units. It is equivalent to about 1/4 pound in the English system of units. Quantity m is the mass measured in kilograms. The prefactor $\frac{1}{2\pi}$ can be calculated from the fact that π is approximately equal to 3.14159. Thus, the prefactor is the number 0.159. Insight into Eq. (3.1) appears in Appendix F.

2. *Shape*: A mode of vibration has a shape. In the case of a mass hanging from a spring, the mode shape is simply the up and down motion of the spring. That is pretty obvious. In the future we will encounter musical systems with modal shapes that are not so obvious.
3. *Amplitude*: If we put a lot of energy into a mode of vibration, its amplitude will be large. But the amplitude is not a "property" of a mode. It only says how much action there is in the mode.

A mode is rather like your bank account. You have an account, and it exists. It has your name on it, and it has a number, just like the frequency of a mode. You can put a lot of money into your bank account or you can take money out of it. Whether the "amplitude" of money in your account is large or small, it is still your account. That is its property.

Fig. 3.2 A spring and mass system with two modes of vibration. Motion of the masses in (1) low- and (2) high-frequency modes are shown by *arrows*

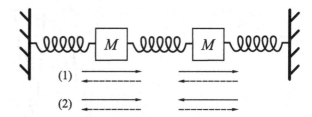

3.3   Multimode Systems

The one-dimensional spring and mass system has one mode of vibration. Most physical systems, including musical systems, have more than one mode. Using springs and masses we can construct a one-dimensional system with two modes of vibration. Figure 3.2 shows such a system. By saying that the system is one-dimensional, we mean that the masses are required to move along a straight line path.

A system of two masses and two springs has two modes of vibration, which means that there are two separate and distinct natural frequencies for this system, one for each mode. The shapes of the two modes are quite different. In the mode with the lower frequency (1), the mode shape has the two masses moving together in lock step. That means that the spring in the middle is not stretched or compressed. In the mode with the higher frequency (2), the two masses move contrary to one another. When the first mass moves to the left, the second mass moves to the right and vice versa. Those mode shapes are shown by the arrows in Fig. 3.2.

Possibly you are now doing a bit of *inductive reasoning* and thinking that if a spring and mass system with one mass has one mode, and a system with two masses has two modes, then a system with three masses would have three modes. If this is what you are thinking, you would be right. And so on it goes with four and five masses, etc.

Systems of springs and masses are discrete systems. Most familiar vibrating objects, like drumheads, guitar strings, and window panes, are continuous systems. Such systems have many modes of vibration but sometimes only a few of them are important. It is often possible to identify the *effective springs* and the *effective masses* of such continuous system. The tuning fork provides an example.

3.4   The Tuning Fork

The tuning fork is a vibrating system with several modes of vibration, but one is much more important than the others.

1. *Main mode*: The main mode of a tuning fork has a frequency that is stamped on the fork so that the fork becomes a reference for that frequency. It makes

Fig. 3.3 (a) The main mode of a tuning fork. (b) The first clang mode

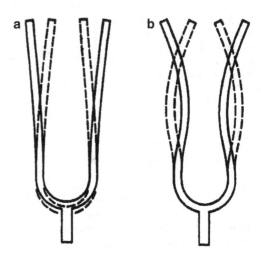

a good reference because it is extremely stable. We actually know how Mozart had his piano tuned because we have his tuning fork. We are rather sure that the frequency of the main mode of this tuning fork has not changed since Mozart's time (1756–1791). A few hundred years should not make any difference to a tuning fork.

2. *Other modes*: There are other modes of the tuning fork. Their frequencies are higher than the main mode. When a tuning fork is struck, these frequencies can also be heard. One of them, the "first clang mode," is particularly evident.

3. *Mode addition*: When a physical system vibrates in several modes at once, the resulting vibration is just the sum of the individual modes. Physicists sometimes call this addition property "superposition." The result of the adding is that the main mode and the clang modes of the tuning fork coexist, and a listener hears all the modes.

Example:

   Suppose we have a tuning fork with a main mode having a frequency of 256 Hz and with a first clang mode having a frequency of 1,997 Hz. Suppose further that we observe the fork when the amplitude of the main mode is 1 mm and the amplitude of the first clang mode is one-third of that or 0.33 mm. Algebraically, the vibration is given by the sum of two sine waves with the correct frequencies and amplitudes. It is

$$x(t) = 1.0 \sin(360 \cdot 256\,t) + 0.33 \sin(360 \cdot 1997\,t) \qquad (3.2)$$

   Figure 3.4 shows how this addition works. Part (a) shows vibration in the main mode with a frequency of 256 Hz and an amplitude of 1 mm. Part (b) shows vibration in the first clang mode with a frequency of 1,997 Hz and an

---

amplitude of 0.33 mm. The sum of Parts (a) and (b) is given in Part (c). If we put a microphone a few centimeters away from the tuning fork we could capture that sum.

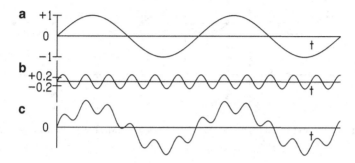

Fig. 3.4 Waveforms (**a**) and (**b**) are added together to make (**c**). The waveforms are functions of time, t

4. *Damping*: The clang modes do not interfere much with the use of the tuning fork because these modes are rapidly damped. Their amplitudes become small soon after the fork is struck. By contrast, the main mode lasts a long time—many minutes! One minute after the fork is struck, only the main mode can be heard. The fork effectively becomes a single mode system, just like a single spring and mass.
5. *Mode shapes*: The shapes of the main mode and the first clang mode appear in Fig. 3.3. They are drawn the way we always draw modes of vibration, with a solid line indicating the maximum displacement in one direction and a dashed line indicating the displacement half a period later in time, when the vibration has a maximum in the opposite direction.

3.4.1 Wave Addition Example

Adding two waves, as in Fig. 3.4, means that at every point in time the values of the two waves along the vertical axes are added together to get the sum. An example is given in Fig. 3.5.

Helpful rules for this kind of point-by-point addition are:

1. When one of the two waves is zero, the sum must be equal to the value of the other wave. This rule is illustrated by filled circles in the figure.
2. When both waves are positive the sum is even more positive, and when both waves are negative the sum is even more negative. These rules are illustrated by the square and triangle symbols.

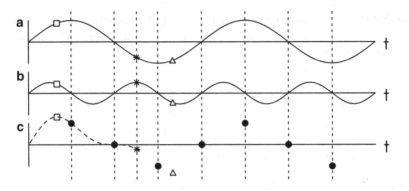

Fig. 3.5 Waves (**a**) and (**b**) are added together to get (**c**). Most of (**c**) is missing—a creative opportunity for the reader

3. When the waves are of opposite sign, the sum tends to be small. This rule is illustrated by the star symbols.
4. To eyeball an accurate plot one can draw a series of vertical lines (fixed points in time) and plot a series of dots that show the sum of the waves. Then by connecting the dots one gets the summed wave, as shown by the dashed line in part (c) of the figure.

3.5   The Spectrum

The waveform representation of a vibration (shown in Fig. 3.4c) is a function of time. An alternative representation is the spectral representation. It describes the vibration as a function of frequency. The amplitude spectrum shows the amplitude of each mode plotted against the frequency of the mode.

Remember that the vibration shown in Fig. 3.4c was the sum of a 256 Hz sine vibration with an amplitude of 1 mm and a 1,997 Hz sine vibration with an amplitude of 0.33 mm. The amplitude spectrum then looks like Fig. 3.6.

Notice how the heights of the waveforms in parts (a) and (b) of Fig. 3.4 directly translate into the representations of amplitudes in Fig. 3.6.

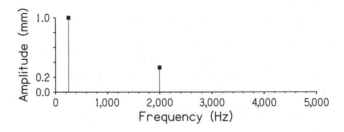

Fig. 3.6 The spectrum of the tuning fork has two lines, one for each mode

3.6 Resonance

The word "resonance" is often used (and misused!) in popular acoustical discussions. It is a word that has a technical meaning, and we shall deal with that technical definition right now.

The concept of resonance requires that there be two physical systems, a *driven* system and a *driving* system. As you can imagine, the driving system feeds energy into the driven system. For example, for the acoustic guitar, the driving system is a vibrating string, and the driven system is the guitar body. The body, with its top plate and sound hole, is responsible for radiating the sound. The guitar body vibrates because energy is fed into it from the string.

The concept of resonance requires that the driven system have some modes of vibration. Therefore, there are special frequencies where this driven system naturally vibrates if once started and left alone. Resonance occurs when the driving system attempts to drive the driven system at one of these special frequencies. When the driven system is driven at a frequency that it likes, the driven system responds enthusiastically and the amplitude of the resulting vibration can become huge.

3.6.1 A Wet Example

Imagine a bathtub that is half full of water. You are kneeling next to the tub and your goal is to slop a lot of water out of the tub and onto the bathroom floor. (The rules of this game are that you can only put one hand into the tub.) You could splash around randomly in the tub and get some water out, but if you really want to slop a *lot of water* you would use the concept of resonance.

A little experimenting would quickly show you that there is a mode of vibration of water in the tub, sloshing from one end to the other. This mode has a particular natural frequency. Your best slopping strategy is to use the palm of your hand to drive this mode of vibration—moving your hand back and forth (sine-wave motion) in the tub at this frequency. It would not take you long to learn what the natural frequency is and to move your hand at the correct frequency to match the natural frequency. You would see a resonant behavior develop as the amplitude of the water motion became larger and larger. Moving your hand at any other frequency would be less effective. Very soon you and the floor would be all wet.

3.6.2 Breaking Glassware

A popular motif in comedy films has a soprano singer, or other source of intense sound, breaking all the glassware in the room. Even bottles of gin miraculously

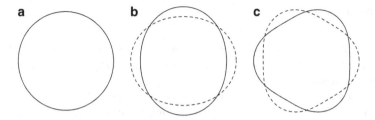

Fig. 3.7 (**a**) A quiet goblet, top view, has no vibration. (**b**) The main mode of vibration has four nodes. (**c**) The second mode has six nodes. The amplitudes are greatly exaggerated in the figure

burst when exposed to this intense sound. The fact is, it isn't that easy to break things with sound. However, it can be done. We can break a crystal goblet.

A crystal goblet is a continuous system with many modes of vibration, but it has a main mode of vibration that is particularly important. This is the mode that is excited if you ding the goblet with your fingernail. It is the mode that is continuously excited if you run a wet finger around the rim. The shape of the mode looks like Fig. 3.7b.

As you can tell from dinging the goblet, this mode of vibration takes a long time to decay away. A long decay time indicates that the resonance is "sharp," meaning that this mode of vibration can be excited by a sine-wave driver only if the frequency of the driver is very close to the natural frequency of the mode. In order to break the goblet, we cause it to have a very large amplitude in this mode of vibration by driving it with a loudspeaker at precisely the natural frequency of this mode. In other words, we use the phenomenon of resonance. We say that we are driving the goblet at its resonance frequency. The goblet is so happy to vibrate at this frequency that it responds with a large amplitude and finally vibrates itself to death.

3.6.3 Sympathetic Strings

Several Asian musical instruments employ the principle of resonance by adding *sympathetic strings* to the instrument, in addition to the main strings that are actually played. The sympathetic strings are tuned to frequencies that will occur in the piece of music to be played. The term "sympathetic" is apt because these strings might be said to vibrate "in sympathy" with the tones played on the main strings. The best known of these instruments is the Indian **sitar** (Fig. 3.8) a fretted, plucked string instrument which has about 20 strings, only six or seven of which are played. The **sarod** is another Indian instrument without frets, and having four or five melody strings, two drone strings, and about ten sympathetic, resonating strings. The **rubab** is a similar instrument from Afghanistan with three melody strings, three drones, and about ten sympathetic strings.

Fig. 3.8 The sitar. This instrument appears to have 18 strings

An alternative description of sympathetic strings would be to call them "resonators." A *resonator* is always a driven system, specifically designed to enhance the vibration or sound of the driving system.

Exercises

Exercise 1, Natural vibes

Find the natural frequency of a spring and mass system where the spring has a stiffness of 100 N/m and the mass is 250 g (0.25 kg).

Exercise 2, Filing the fork

Think about the main mode of the tuning fork. What part of the tuning fork acts as the mass? What part acts as the spring? What happens to the frequency of the tuning fork if you file away metal from the ends of the tines? What happens to the frequency if you file away metal near the junction of the two tines?

The fact that we can think about a continuous system like a tuning fork in terms of a single mass and spring shows the conceptual power of modal analysis. But which part of the system plays the role of the mass and which part plays the role of the spring depend on the shape of the mode we are considering.

Exercise 3, Lifetime

Look at the damped vibrations in Fig. 3.1. If the frequency of vibration is 200 Hz, how long does it take for the wave to die away? This exercise requires you to find your own definition for "die away." How much decrease is needed in order to say that the vibration has disappeared?

Exercise 4, A simple spectrum

Draw the spectrum for a 1,000-Hz sine tone.

Exercise 5, More complex spectra

Draw the spectra for the following waves:

(a) $x(t) = 1.3 \sin(360 \cdot 575\,t) + 0.8 \sin(360 \cdot 1200\,t)$
(b) $x(t) = 0.5 \sin(360 \cdot 100\,t) + 2.2 \sin(360 \cdot 780\,t) + 0.8 \sin(360 \cdot 1650\,t)$

Exercise 6, Resonance—an anthropomorphism

If your friend makes a statement and you agree with the statement, it might be said that the two of you are in resonance. Explain how this metaphor is an extension of the concept of resonance for physical systems as explained in this chapter.

Exercise 7, Playground Resonance

(a) Explain how the idea of resonance is used to great effect when pushing a child in a swing.
(b) Is the swing example an exact example of resonance?

Exercise 8, Square root dependence

According to Eq. (3.1), the frequency of a spring and mass system depends on the ratio of spring stiffness to mass. However, it does not depend on this ratio in a linear way. It is a square-root dependence.

(a) What happens to the frequency if you change the system by doubling the stiffness and doubling the mass, both at once?

(b) What happens to the frequency if you double the mass?

Exercise 9, Adding waveforms

Figure 3.9 shows two waveforms, (a) and (b). They show displacement as a function of time for two simultaneous modes of vibration. (A) Add these waveforms—point by point—to find the resulting vibration pattern. (B) If the frequency of the wave in part (a) is 100 Hz, what is the frequency of the wave in part (b)?

Exercise 10, Resonance?

Which of the following are examples of resonance?

(a) You strike a Tibetan singing bowl and it vibrates in several of its modes.
(b) You run a wet finger around the rim of a crystal goblet and it vibrates in its main mode.
(c) You strike a tuning fork and it ultimately vibrates only in its main mode.
(d) You bow a violin string and it vibrates at its natural frequency.

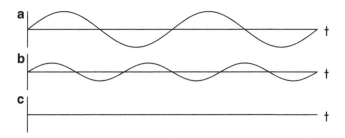

Fig. 3.9 Waveforms to be added for Exercise 9

Fig. 3.10 Modes of a
kalimba tine for Exercise 11.
The tine is fixed in the bridge
at the left side

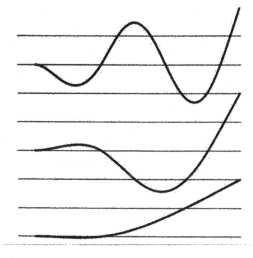

Exercise 11, Kalimba tine

The kalimba is a westernized version of the African mbira or "thumb piano." It
consists of metal tines mounted on a wooden box. The performer uses his thumbs
to pluck the tines and the box acts as a resonator. Figure 3.10 shows the first three
modes of vibration of a kalimba tine, as calculated by David Chapman and published
in 2012 in the Journal of the Acoustical Society of America. Which mode do you
think has the highest frequency? Which has the lowest?

Exercise 12, Kalimba box

By itself, the kalimba tine does not radiate sound effectively. The kalimba box
solves that problem. The box is hollow and has a sound hole like a guitar. The
kalimba box vibrates, as forced by the tines, and its large surface area and sound
hole become effective radiators of the sound. The box also has resonances because
of the modes of vibration of air inside the box. Based on your experience with
hollow objects, do you think that the resonance frequencies become higher or lower
if the box is made bigger and its volume increases?

♠

Chapter 4
Instrumentation

Acoustics is an ancient science. In the fifth century BC, Pythagoras made the science of musical tones into an entire philosophy. In renaissance times Galileo and Descartes developed the mathematics of musical tones, and Isaac Newton tried (but failed) to calculate the speed of sound in air. The nineteenth century saw the beginning of truly serious study of human hearing—the anatomy, the physiology, and the psychology. Herman von Helmholtz led the way.

In the twentieth century the study of acoustics got two tremendous boosts. The first boost arrived in the first half of the century with the introduction of electronics and the invention of acoustical instrumentation. For the first time, scientists had a tool for studying sound that was fast enough to keep up with the full frequency range of sound itself. The second boost arrived in the second half of the century with the widespread availability of computers and digital signal processing. After these two revolutions, nothing in the ancient science of acoustics will ever be the same again.

This chapter is about electronic instrumentation. It deals with the subject of transducers and with four instruments: the oscilloscope, the spectrum analyzer, the frequency counter, and the function generator.

4.1 Transducers

A transducer is a device that converts a signal from one form to another. If you want to make an audio recording, you use a microphone. A microphone is a transducer. As indicated by Fig. 4.1, it converts a signal from the acoustical domain, where it is a pressure waveform transmitted through the air, to the electrical domain, where is it a voltage waveform transmitted through wires.

To do the reverse process, converting an electrical signal into an acoustical pressure waveform in the air, you use another transducer—a loudspeaker or a set of headphones.

W.M. Hartmann, *Principles of Musical Acoustics*, Undergraduate Lecture Notes
in Physics, DOI 10.1007/978-1-4614-6786-1_4,
© Springer Science+Business Media New York 2013

Fig. 4.1 A microphone
converts a signal from an
acoustical pressure wave (p)
to an electrical voltage (v). A
loudspeaker does the reverse

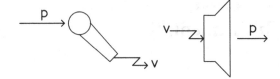

The microphone and loudspeaker are known as analog devices. They are called so because the output of the device is analogous to the input. For instance, if a microphone receives a sound pressure waveform of a particular shape, then the output from the microphone ought to be a voltage waveform that preserves the details of that shape at every instant in time. If the output does not exactly represent those details, then the microphone has distorted the signal.

A well-behaved transducer that does not distort the signal is *linear.* In a linear transducer, a change in the input to the devices causes a proportional change in the output. If from one instant of time to the next the input changes by a factor of 1.27 (becomes 1.27 times larger), then the output must change by a factor of 1.27. If it changes by a factor of 1.26 instead of 1.27 there is distortion.

Linearity can be shown on a graph that describes the relationship between the input and the output. The graph must be a straight line as in Fig. 4.2a. That is the meaning of the word "linear." The idea of linearity also normally assumes that the straight line goes through the origin so that the output is zero when the input is zero. The operation of microphones, loudspeakers, and other transducers is further discussed in Chap. 16.

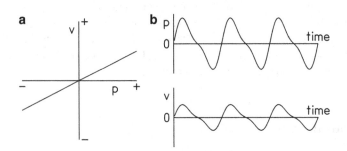

Fig. 4.2 (**a**) The input/output relationship of a linear transducer is a straight line. For a microphone, the output in volts (v) is linearly related to the input in pressure units (p). Positive and negative axes are shown. (**b**) The input pressure waveform is transduced by the microphone into an analogous voltage waveform. Because the slope of the line in (**a**) is shallow, the voltage (v) in part (**b**) is shown smaller than the pressure (p), but the shape of the pressure waveform is preserved in the voltage waveform

4.2 The Oscilloscope

The oscilloscope (or "'scope") is an instrument that displays waveforms as functions of time. The horizontal axis of the display represents time and the vertical axis represents a voltage. The voltage may come from a microphone, so that in the end the oscilloscope can show a sound wave and its dependence on time.

There are two types of 'scopes available, analog and digital. The analog 'scope may use a cathode ray tube (CRT) for the display, as shown in Fig. 4.3. CRT displays were formerly the standard displays in television sets and computer monitors. The digital 'scope uses either a CRT display or a liquid crystal display (LCD). LCD screens are also found in laptop computers, ipads, graphing calculators, and cell phones.

4.2.1 Analog 'Scope and CRT Display

A "cathode ray" is actually a beam of electrons shot out from an electron gun toward the screen in the front of the tube. The screen is coated with a fluorescent material that glows briefly when it is hit with a high-speed electron. The electron beam from the gun speeds down the neck of the tube and hits the screen directly in the center. Then if you look at the screen you will see a single small dot, right in the middle.

To make a useful picture we need to move the dot from left to right and up and down on the screen. We need to move it so rapidly that we don't see a moving dot—we see a picture instead. The dot is moved electronically by horizontal and vertical deflection mechanisms that deflect the electron beam from its straight-line path. This is done using the charged properties of the electron. Deflection plates are placed in pairs—top and bottom, and left and right—within the CRT. These deflection plates are given either a positive or negative charge to deflect the electron in the desired direction. The oscilloscope gives the user great control over the way the dot is moved.

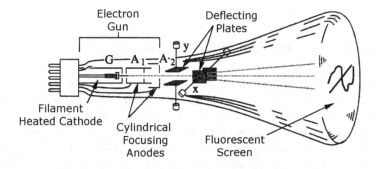

Fig. 4.3 A cathode-ray tube is the display for 'scopes without digital memory

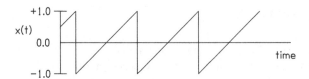

Fig. 4.4 The sawtooth waveform always increases in the sense that it never spends any time decreasing. When it controls the horizontal position of the dot, the dot always moves from *left* to *right*. A dot controlled by the sawtooth wave shown here would start to the *right of center*, about half way to the *right-hand edge*. It would move to the *right*, snap back to the *left-hand edge* and move to the *right* again three times

Horizontal Deflection of the Analog Oscilloscope We would like the analog oscilloscope to draw a waveform as a function of time. That means that the horizontal motion of the dot must sweep from left to right proportional to time. The way to make this happen is to use a sawtooth waveform for the horizontal deflection circuit. Figure 4.4 shows this waveform always moving from a value of −1 (left) to a value of +1 (right). The sawtooth is a straight-line function of time, and so the horizontal motion is proportional to time. There are brief instants when the sawtooth jumps back from +1 to −1, but these jumps take place almost instantaneously and they leave no trace on the screen. The rate of the sawtooth, the number of sawtooth cycles per second, is something that the user can freely adjust to get a good display. To see a signal that has a high frequency requires that the sawtooth also be rapid. That is a good reason for using CRTs with deflection plates—the fastest known way to move an electron beam.

Vertical Deflection of the Analog Oscilloscope The vertical deflection of the dot on the screen in the analog 'scope is controlled by the signal that we want to display. For instance, if we want to examine the sound radiation of a tuning fork, we put a microphone near the fork to get an analogous voltage. Then this voltage goes to the *vertical amplifier* of the oscilloscope to adjust its amplitude to a convenient value. The output is thus analogous to the input, hence the name analog oscilloscope. In the end, the 'scope displays the microphone signal (y-axis) as a function of time (x-axis).

4.2.2 Digital 'Scopes and Liquid Crystal Displays

The difference between analog and digital oscilloscopes is that a digital 'scope "digitizes" the input and stores it in digital memory before it is displayed. A digital scope has a converter, like the sound card on a computer, to convert analog signals to digital signals. Once the signal is in digital form it can easily be manipulated and displayed. Because the digital signal is stored in the oscilloscope's memory the display doesn't have to be as fast as the display in an analog 'scope. The digital display doesn't have to keep up with the signal in real time. This gives the digital 'scope display options such as liquid crystal displays.

Just as cathode-ray tubes are being replaced by liquid crystal and plasma displays in televisions, so too CRTs are being replaced by LCDs in oscilloscopes. In contrast to the analog CRT, which draws continuous line, the LCD uses a series of discrete pixels. Each pixel is created from a liquid crystal. A liquid crystal is made from molecules that are twisted unless an electric signal is applied to them. Applying an electric signal untwists the molecules. The LCD screen is made with polarizing glass such that when the crystals are twisted they let the light through, and when they are untwisted they block the illuminating light. In this way the LCD screen creates an image by using the electric signal from the input source to determine which pixels to transmit light and which pixels to block out light.

4.2.3   Beyond the Basic Oscilloscope

Important refinements to the basic oscilloscope include the following:

(a) *Triggered sweep*: A circuit that synchronizes the start of the horizontal sweep with the waveform being displayed. With triggered sweep, subsequent traces of a periodic waveform are identical on the screen. That's important because, at any one time, there are many traces visible on the screen. If the traces are not synchronized, the image is a jumble.

(b) *Dual trace*: The ability to display two different waveforms on the same screen at the same time. The two waveforms are drawn simultaneously, causing them to be perfectly aligned in time. The dual-trace feature allows the user to compare waveforms. For instance, one can compare the input and the output of a device to determine the effect of the device.

(c) *Digital memory*: The ability to store a waveform. In a conventional analog 'scope, when the signal dies away the amplitude on the display dies away as well. How could it do anything else? But suppose that we were able to capture a waveform in memory and keep on playing it back to the vertical deflection circuit. Then we could capture a transient event, like a drum beat, and watch it for as long as we want. That's a big advantage of the digital 'scope.

4.3   The Spectrum Analyzer

The spectrum analyzer is like the oscilloscope in that it displays a picture on a screen, but that is the end of the resemblance. The spectrum analyzer displays amplitudes as a function of frequency. What a difference! In the oscilloscope the horizontal axis is time; in the spectrum analyzer it is frequency. In the oscilloscope

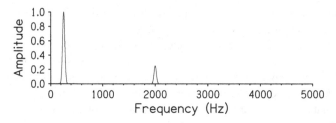

Fig. 4.5 If a tuning fork has a main mode at 256 Hz and a clang mode at 1,997 Hz, then the spectrum analyzer will display two peaks. The heights of the peaks indicate the relative amplitudes in those two modes

the vertical axis is the signal itself; in the spectrum analyzer it is the amplitude of the component whose frequency corresponds to the value on the horizontal axis.

Obviously the spectrum analyzer is just what we need if we want to analyze a signal into its sine wave components. For instance, if we put in a signal from a tuning fork shortly after it is struck, then the spectrum analyzer will indicate the presence of amplitude at the main frequency and at the frequency of the clang mode. Ideally the display would be two vertical lines like Fig. 3.6. Instead, these lines always have a certain width causing the vertical spikes to become peaks, as shown in Fig. 4.5.

The analysis performed by the spectrum analyzer is far more complicated than simply displaying the signal on an oscilloscope. It takes a certain amount of time to do the analysis. In a swept-frequency analyzer, a trace goes from left to right on the screen as the instrument makes its analysis, frequency by frequency. The analysis is not complete until the trace reaches the right-hand edge. The signal has to last for the entire duration of the analysis. If the signal changes during the course of analysis, the analysis will be inaccurate. Therefore, the swept-frequency analyzer would not really be appropriate to study a decaying signal like the tone of a tuning fork. Once again, digital electronics can help. A digital spectrum analyzer can store a brief signal in memory and perform the frequency analysis later. Alternatively, the digital spectrum analyzer can make successive analyses of a long signal, refining its measurement on each pass.

The amplitude displayed on the vertical axis of the CRT display or LCD is not exactly like the display shown in Fig. 4.5. Instead, it is normally displayed in decibels. The decibel transformation of amplitude allows the analyzer to display components of greatly different amplitude all on the same graph on the screen. We will study this decibel transformation in Chap. 10.

4.4   The Frequency Counter

The frequency counter does not have a screen like an oscilloscope. It just has a numerical readout with half a dozen digits. The readout indicates the frequency (in Hz) of the signal that is put into the counter. An obvious problem is that a signal may have many different frequencies—more about that later (Fig. 4.6).

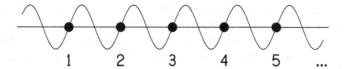

Fig. 4.6 The frequency counter counts positive-going zero crossings in a signal that comes through the gate

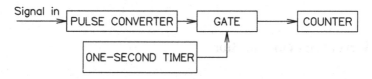

Fig. 4.7 Block diagram showing the functional components of a frequency counter

The frequency counter works by mindlessly counting the occurrence of a particular feature in the signal voltage. It is common to count positive-going zero crossings. You know that a sine signal has one positive-going zero crossing per cycle. So does a sawtooth signal (Fig. 4.4). Therefore, by counting positive-going zero crossings, the counter is counting the number of cycles. The number of cycles is displayed on the numerical readout. It counts $001, 002, 003, 004, \ldots$ As shown in Fig. 4.7, the next ingredient in the counter is an electronic gate that tells the counter when to count. The gate is precisely 1 s long. Therefore, the counter reads out the number of cycles that occur in 1 s. Wow! That is exactly what we mean by frequency. Now we have a frequency counter.

When we use the frequency counter, we are making several assumptions. We assume that the signal does not change during the 1-s gate interval required to count cycles. We also assume that there is only one positive-going zero crossing per cycle. If the signal is a sine wave then that assumption is true, but if the signal has many frequencies the assumption may or may not be true.

4.5 The Function Generator

The oscilloscope, spectrum analyzer, and frequency counter are all instruments of analysis. We put a signal into them and we get out a measurement or display that gives us information about the signal. The *function generator* is the reverse. The function generator creates a signal; it creates the signal in electronic form. The user must specify the frequency and the amplitude by some means, perhaps by knobs on the front panel of the instrument, perhaps by computer control. The user can also specify the waveform, e.g., sine wave, sawtooth wave, square wave, or triangle wave. Such simple waveforms can be used to test audio and acoustical systems. They also are at the basis of electronically synthesized music, where they are called "oscillators."

Digital function generators are sometimes called "waveform synthesizers." A waveform synthesizer allows the user to generate complicated waveforms of any desired shape, even transient waveforms. Engineers at Ford in Dearborn, Michigan, use huge transducers under the wheels of test vehicles to subject these vehicles to realistic (and unrealistic) vibrations. The transducers are driven by waveform synthesizers that simulate bumpy roads, dips, potholes, etc.—anything that will torture a car and reveal weaknesses.

4.6 Virtual Instrumentation

Can a special-purpose instrument, such as an oscilloscope or a spectrum analyzer, be replaced by a general purpose computer? What features of the instrument would the computer need to accomplish this? It would need a way to acquire a signal (an input). It would need a way to process and represent the signal as selected by the user. It would need a way to display the result. The computer has these features. It has a sound card that can input signals. It has a memory and arithmetic capability that can follow a program. It has a monitor screen for display. Computer programs that emulate the functions of special-purpose instruments are generically known as virtual instruments. It is usual for the display to simulate the screen of the real special-purpose instrument. The display also depicts the knobs, pushbuttons, dials, and indicators that are typical of the real instrumentation. The user manipulates these controls with a mouse. Good virtual instrumentation can rival the quality and ease of use found in real instrumentation. An advantage of the real instrumentation is that real instrumentation is often used to measure signals that have been generated by a computer. The real instruments offer an independent confirmation.

Exercises

Exercise 1, The telephone
   How many acoustical transducers are in a telephone handset? What are they?

Exercise 2, Fooling the frequency counter
   On the graph in Fig. 4.8 draw a periodic signal that has two positive-going zero crossings per cycle. Explain why a frequency counter would display a frequency that is two times larger than the fundamental frequency of the signal. (A factor of two in frequency is called an "octave.")

Exercise 3, All things come to him who waits.
   A frequency counter displays a frequency as an integer number of cycles per second because it counts discrete events. For instance, it might display 262 Hz. Suppose you wanted better accuracy. Suppose you wanted to know that the frequency was really 262.3 Hz. What could you do?

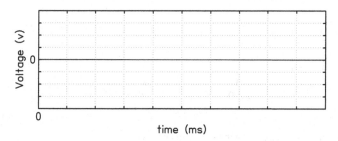

Fig. 4.8 Your signal to fool the counter with two positive-going zero crossings per cycle

Exercise 4, Make a list

Microphones and loudspeakers are transducers. Name some other transducers.

Exercise 5, Setting the sweep rate

Suppose you want to display two cycles of a 500-Hz sine tone on an oscilloscope. What should you chose for the period of the sawtooth sweep on the horizontal axis?

Exercise 6, The dual-trace oscilloscope

What is so important about a dual-trace oscilloscope? In what way does it have more than twice the capability of a single-trace 'scope?

Exercise 7, Oscilloscope deflection

Recall that like charges repel and opposite charges attract. The analog oscilloscope deflects the electron beam horizontally using charged plates on the right and left side of the CRT. If you want to deflect an electron (negative charge) to the left side of the oscilloscope screen, which plate would be positively charged, and which plate would be negatively charged?

Exercise 8, Reading the frequency from the oscilloscope

The oscilloscope has ten divisions in the horizontal direction. You observe that the sweep rate control is set to "0.5 ms/div." (Translation: 0.5 ms per horizontal division on the screen.) You see exactly four cycles of a sine tone on the screen. What is the frequency of the tone?

Exercise 9, Estimating the frequency

The oscilloscope is set up as in Exercise 8. You observe that a single cycle of the tone covers slightly more than 2 divisions. Estimate the frequency of the tone.

Exercise 10, Spectrum analyzer and oscilloscope

A signal is sent to both the spectrum analyzer and the oscilloscope. The spectrum analyzer display looks like Fig. 4.9. Explain why the oscilloscope tracing looks like Fig. 4.10.

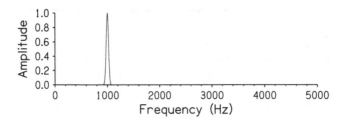

Fig. 4.9 Signal for Exercise 10 on the spectrum analyzer

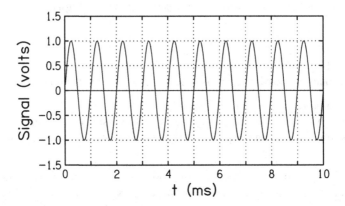

Fig. 4.10 Signal for Exercise 10 on the oscilloscope

Exercise 11, Oscilloscope grid

The oscilloscope grid in Fig. 4.10 has divisions spaced by 1 cm. How many volts per centimeter on the vertical scale? How many seconds per centimeter on the horizontal scale?

♠

Chapter 5
Sound Waves

Recall that the first step in an acoustical event is a vibration of some kind and the second step is the propagation of that vibration. The opening chapters of this book introduced vibrations. Now it is time to look at that second step. Vibrations are propagated by means of sound waves. Therefore, this propagation raises the topic of waves themselves. As you know, there are many different kinds of waves—radio waves, light waves, X-rays, and waves on the surface of a lake. Sound waves are another important example. The different waves have some common features. They all have some frequency (or frequencies). They all can be characterized by a direction of propagation and a speed of propagation, though the speeds are different for different kinds of waves.

In general terms, a wave can be thought of as a physical means of propagating energy, momentum, and information without transporting mass. If a friend throws a ball to you, energy and momentum are propagated, and so is the mass of the ball. By contrast, if a friend subjects you to a booming bass sound, energy and momentum are again propagated (your chest walls may be temporarily deformed) but no mass is propagated. The sound comes to you as a wave. If a friend sends you a letter, you get the information and the mass of the letter. By contrast, if a friend contacts you by cell phone, you get information but no mass. The information is propagated by a radio wave.

We are particularly interested in sound waves. Sound waves can travel through solids, liquids, and gasses. What is normally most important for us as listeners is the propagation of sound waves in air (Fig. 5.1).

To begin the study of sound waves in air, we start with a brief detour into the water. When you dive down into the water you feel a pressure on your eardrums. That is because of the weight of the water on top of you. The weight of the water is due to the earth's gravity, and this weight leads to pressure.

When you are on the surface of the earth, you are under a sea of air. Because of gravity, the air has weight and leads to pressure. It is known as *standard atmospheric pressure*. A pressure has units of a force per unit area. In the metric system of units, force is measured in Newtons and area is measured in square meters, and so pressure

W.M. Hartmann, *Principles of Musical Acoustics*, Undergraduate Lecture Notes
in Physics, DOI 10.1007/978-1-4614-6786-1_5,
© Springer Science+Business Media New York 2013

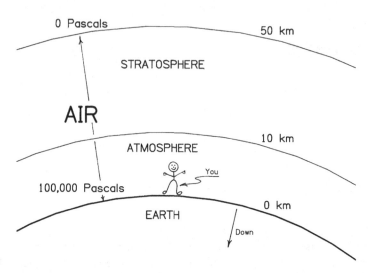

Fig. 5.1 You live under a sea of air, which creates atmospheric pressure (you may have noticed that this figure is not to scale

is measured in units of Newtons per square meter, otherwise known as Pascals. One Pascal (Pa) is equal to a pressure of 1 N/m^2.

Atmospheric pressure is about 10^5 or 100,000 Pa. That looks like a rather large number. There are three reasons that it looks large. The first reason is that a square meter is a large surface area. (Atmospheric pressure is only 10 N per square *centimeter*.) A second reason is that a Newton is not really very heavy—1/4 pound. But the most important reason that atmospheric pressure looks large is that it *really is* large. Atmospheric pressure is equivalent to 14.7 pounds per square inch. To make that meaningful, understand that this means 14.7 pounds of force on every square inch of your body. Wow! You experience that pressure all the time... not just on Mondays. The only reason that you are not squashed by that pressure is that your insides are pressurized too. Your entire vascular system is pressurized equivalently, and you are unaware of the great weight that is upon you. You gain a little insight into this matter when you go up in an airplane or an elevator where there is slightly less atmosphere above you and consequently less air pressure. Your ears may "pop."

A sound wave is a small disturbance in the air pressure. Where a sound wave has a peak, the air pressure is slightly greater than atmospheric pressure. Where a sound wave has a valley, the air pressure is slightly less than atmospheric pressure. The change in air pressure caused by a sound wave is very tiny.

Look at Fig. 5.2. Those huge numbers on the vertical axis show pressure in Pascals. Standard atmospheric pressure is 101,325 Pa. The sound wave makes positive and negative excursions around this value. Its amplitude is only 28 Pa. Probably you are thinking that this does not look like much of a wave. Its amplitude is less than 3 one-hundredths of a percent of the atmospheric pressure. In fact, however, this wave is huge. It is a 120-dB sine tone, and it is so intense that it

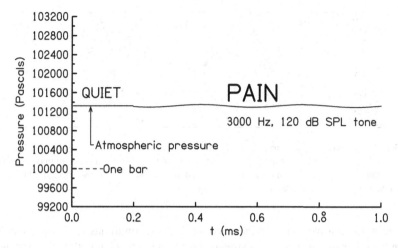

Fig. 5.2 A sound wave is a variation in air pressure, above and below the ambient atmospheric pressure of 101,325 Pa. This figure starts with a flat line. The flat line corresponds to constant atmospheric pressure, namely complete silence. After 0.2 ms a sine wave is turned on. The sine wave is intense enough to cause pain in human listeners (a pressure of exactly 100,000 Pa is known as one "bar"—same word root as barometer)

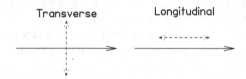

Fig. 5.3 Two waves propagate from left to right as shown by the *solid arrows*. The disturbance (*dashed*) is perpendicular to the propagation for a transverse wave. The disturbance is parallel to the propagation for a longitudinal wave

causes pain to human listeners. A quiet wave, corresponding to the threshold of hearing, has an amplitude that is one million times smaller. It should be clear that sound waves correspond to very miniscule changes in air pressure.

5.1 Polarization

There are two directions associated with any wave, a direction of propagation and a direction of displacement (Fig. 5.3). In a *transverse* wave the direction of displacement is perpendicular to the direction of propagation. If one end of a rope is tied to a tree and you shake the other end, you propagate a wave along the rope. The length of the rope defines the direction of *propagation*. The direction of *displacement* of the rope is either up and down or side to side. Both of these are perpendicular to the direction of propagation, and so both are transverse waves.

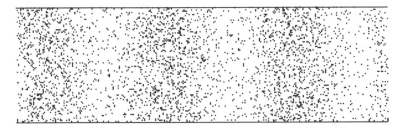

Fig. 5.4 A longitudinal wave propagates from left to right in a hollow tube filled with air. A snapshot of the wave shows that the molecules of air are close together at a pressure maximum and they are more widely spaced at a pressure valley. About three cycles are shown because the length of the entire figure is about three wavelengths long

A transverse wave can exist in a medium if the medium has some resistance to being bent. For instance, a tightly stretched rope and a steel bar both have resistance to bending and both can support a transverse wave.

In a *longitudinal* wave the direction of displacement is parallel to the direction of propagation. A visible example of a longitudinal wave is the vibration of a slinky along its length. As a wave propagates along a stretched slinky, the displacements of the coils are parallel to the extension of the slinky. Therefore, the displacements are in the same direction as the wave propagation.

A longitudinal wave can exist in a medium if the medium has some resistance to being compressed. For instance, air or water both have resistance to compression and both can support a longitudinal wave.

The reason for bringing up this matter of polarization is that sound waves can be either transverse or longitudinal, but when sound propagates in a gas—like ordinary air—the sound wave is longitudinal, as shown in Fig. 5.4. For instance, if a wave propagates in a north direction, then the molecules of air move along a north-south line. A transverse wave cannot exist in air because air has no resistance to being bent. In terms of the spring and mass analogy, there is no transverse spring.

5.2   The Speed of Sound

The speed of sound in room temperature air is 343.7 m/s. That corresponds to 1,128 ft/s or 769 miles/h.

In fact, the speed of sound depends on air temperature. The higher the temperature, the faster the sound. The dependence is not a very strong dependence. A simple formula that can be used over the range of temperatures that we normally encounter gives the speed of sound as

$$v = 331.7 + 0.6\,T_C, \tag{5.1}$$

where v is the speed of sound in meters per second and T_C is the temperature in Celsius degrees. On the Celsius scale, room temperature is 20°, and that is how we get the speed of sound to be 343.7 m/s at room temperature.

Interestingly, the speed of sound does not depend on the air pressure. On a clear day, when the barometer reads high or on a stormy day when the barometer reads low, the speed of sound is the same. The reason is that the effect of changing air pressure leads to mutually compensating changes in the air. Increasing the air pressure increases the elasticity of air, and this tends to increase the speed of sound. But increasing the air pressure also increases the density of the air, and this tends to decrease the speed of sound. In the end, these two changes in the properties of air cancel one another.

5.2.1 Supersonic Things

Something that is supersonic goes faster than the speed of sound. Some airplanes are supersonic, and their speed is measured in Mach number, which is just a way of saying how many times faster than the speed of sound. For instance, the F16 fighter plane flies at Mach 2, which means twice the speed of sound (this was also the cruising speed of the Concorde passenger plane). Because the speed of sound is 769 miles per hour, you might conclude that the F16 flies at 1,538 miles per hour. That would be close but not quite right. The trick here is that the F16 flies at 50,000 ft. At this high altitude the air is cold, and the speed of sound is only 670 miles per hour. Mach numbers are measured with respect to the ambient air. Therefore, for this plane, Mach 2 is only 1,340 mph.

A supersonic thing makes a shock wave that trails behind it so long as it continues to travel at supersonic speeds. For an aircraft traveling faster than Mach 1 this shock wave is called a sonic boom. It can break windows all along the path below the plane for as long as the plane flies supersonically.

Linearizing a Function

Equation (5.1) for the speed of sound is good and it is bad. It is good because it provides a simple way to calculate the speed of sound for any normal air temperature. It is simple because it is linear in the air temperature, T_C. However, the equation is bad because it misrepresents the true way that the speed of sound actually depends on temperature. In reality, the speed of sound depends on the square root of the absolute temperature, referenced to absolute zero (−273 °C). That dependence is shown in Fig. 5.5 for temperatures ranging from −194 to +300 °C. It is not linear, it is curved.

The shaded region in Fig. 5.5 extends from −18 to 40°C, corresponding to Fahrenheit temperatures from 0 to 104°. That is the temperature range where we normally live. Over that small temperature range a straight line (linear) is a good approximation to the curve. Equation (5.1) represents that straight line approximation to the curve. It is an example of a mathematical technique commonly used in engineering and economics, linearizing a more complicated function to represent behavior over a limited range.

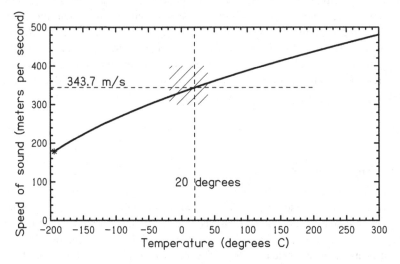

Fig. 5.5 The speed of sound depends on the square root of the absolute temperature, as shown by the *heavy line*, but over a restricted temperature range it is well represented by a *straight line*. (The asterisk at $-194°$ shows where air becomes a liquid.) The *shaded region* indicates the normal range of outdoor air temperatures

5.2.2 *Sound vs Light*

Sound and light are both wave phenomena. However, there are enormous differences between sound waves and light waves. Most important, sound waves require a medium, such as air or water, to propagate. Light waves do not require a medium. Light waves can travel through a vacuum. That is why we can see the sun (visible light) and feel its warmth (infrared light) but we cannot hear the sun.

Sound waves consist of pressure changes in the medium; light waves, including radio waves, consist of changes in an electromagnetic field. Light waves travel at the speed of light (How about that!). The speed of light is 3×10^8 m/s. To compare that with the speed of sound in air, consider that 344 m/s is about 3×10^2 m/s. Evidently light is faster by a factor of 10^6, i.e., it is a million times faster. In the time that it takes a sound wave to go 40 m (130 ft) a light wave can go completely around the earth at the equator (25,000 miles). In the time that it takes sound to go 1 km, a light wave can travel to the Moon and back.

A light wave has a frequency just like a sound wave does. While the frequencies of audible sound are 20–20,000 Hz, the frequencies of visible light range from 4×10^{14} Hz (red) to 7×10^{14} Hz (violet). Also, where sound waves in air are longitudinal, light waves are transverse. The electromagnetic field is perpendicular to the direction of propagation.

5.3   Sound Waves in Space and Time

In previous chapters we have drawn waves as functions of time. Such a wave might represent the vibration of a guitar string at the position of the bridge of the guitar, or it might represent a voltage coming from a microphone. Such waves are functions of one dimension, the dimension of time. When we talk about propagating waves in space, we need to add spatial dimensions to the description of wave motion. Because there are three dimensions of space, we might add as many as three more dimensions.

Often it is possible to consider sound propagation in a single spatial dimension. Then there are two dimensions, one dimension of space and one dimension of time. We can imagine plotting sound pressure as a function of time and a single spatial dimension. To do so requires a three-dimensional graph, as shown in Fig. 5.6.

It is difficult to work with a three-dimensional graph, and most of the time we don't. What we do instead is to decide in advance whether we want to describe the wave in time or in space. If we choose to plot the wave in time, we pick a point in space and draw a graph of the sound pressure as the wave comes by that point. If we chose to plot the wave in space, we pick a point in time and draw a graph of the spatial dependence of the sound pressure. Making a graph at a particular point in time is called a "snapshot." This is a familiar concept. A snapshot photograph shows spatial relationships—e.g., you are in the middle, your friends are to the right and the left—but time is frozen.

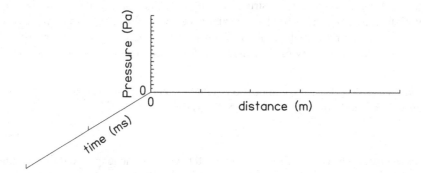

Fig. 5.6 A three-dimensional coordinate system for plotting sound wave as a function of time and one dimension of space. If you have trouble visualizing three mutually perpendicular axes like this, think about the corner of a room, where two walls meet a floor

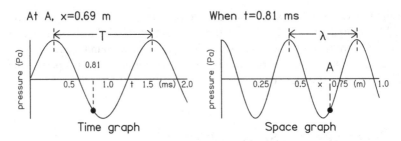

Fig. 5.7 Two views of a 800-Hz sine tone. For the graph on the *left*, we choose a point in space (the 0.69 m mark at point **A**) and we plot pressure as time goes by. We observe that the period is 1.0/800 = 1.25 ms. For the graph on the *right*, we take a snapshot at time $t = 0.81$ ms and we plot the pressure at all points in space at this time. We see that the wavelength is $\lambda = 344/800 = 0.43$ m

The two kinds of plot are shown in Fig. 5.7. The graph on the left shows the pressure as a function of time. The graph on the right shows the pressure as a function of space. The function is a sine wave, which is periodic in time. Its period is T, measured in seconds. Because the time graph is periodic, the space graph is also periodic. Its period has a spatial character. It is called wavelength—lambda (λ)—and it is measured in meters.

There is a simple and important relationship between the frequency of a wave, the wavelength of the wave, and the speed of the wave. The relationship is

$$v = f\lambda \tag{5.2}$$

This equation can be manipulated in different ways with different interpretations. The interpretations are analogies for waves that are intended to make the concepts of frequency and wavelength easier to understand.

Frequency: First, Eq. (5.2) can be used to find the frequency,

$$f = v/\lambda \tag{5.3}$$

Interpretation: There is a rigid structure with a regular spacing like the teeth of a saw or the threads of a screw. The spacing is called λ. When this structure moves past point **A**, the rate at which the teeth or threads come by is equal to the speed of the structure divided by the spacing. That rate is frequency f, as determined by Eq. (5.3).

Period: Of course, Eq. (5.3) can be used to find the period, because the period is the reciprocal of the frequency—just Eq. (5.3) upside down:

$$T = \lambda/v \tag{5.4}$$

Interpretation: The cars on the freight train all have the same length, and that length is λ. The length of time between the passage of freight cars is equal to the length of the cars divided by the speed of the train. That is the period T as determined by Eq. (5.4).

Wavelength: Equation (5.2) can also be written as

$$\lambda = v/f \qquad (5.5)$$

Interpretation: All cars travel on the road at exactly the same speed, v. The speed is measured in miles per hour. The cars are sent out from a starting point at a rate of f cars per hour. After a few hours, the cars are separated by equal distances; that distance is λ, as determined by Eq. (5.5).

Equation (5.5) can be applied directly to Fig. 5.7. There is a sine wave with a frequency of 800 Hz. The speed of sound is 344 m/s, and so the wavelength is $344/800 = 0.43$ m.

It may look as though Eqs. (5.2)–(5.5) require you to memorize four equations. In fact, there are only two, $v = f\lambda$ and $f = 1/T$. The other equations can be derived from these two. Please try to think about it that way. It will help your math skills.

Changing Media Equations (5.3), (5.4), and (5.5) describe the relationship between frequency, wavelength, and the speed of sound. What happens when a sound wave goes from one medium to another so that the speed of sound changes? Somehow those equations need to be satisfied. Does the wavelength change, or does the frequency change, or do they both change? The answer is that the frequency stays the same and the wavelength changes to accommodate the new speed of sound.

When a sound wave travels from air into water, where the speed of sound is five times greater, the wavelength increases by a factor of 5. (To be more complete, when a sound wave in air hits the water, most of the wave is reflected. But that part of the wave that does enter the water has its wavelength increased by a factor of 5.)

5.4   Sound Waves in More Than One Dimension of Space

The section called "Sound waves in space and time" considered only one dimension of space. That description applies well to a tin-can telephone, where sound waves travel on a stretched string, but otherwise you know that there are three spatial dimensions. From any point you can move forward–backward, left–right, or up–down. That makes three. We need to deal with this problem of spatial dimensions.

One-Dimension We begin by trying to justify the previous discussion dealing with only one spatial dimension. First, with a source at one point in space and a receiver at another, there are two points, and they can be connected by a straight line. Arguably that is the most interesting path for a wave, and that path is one dimensional. Thus, a situation that really has three dimensions of space has been reduced to a situation that has only one. A second justification is that a one-dimensional treatment is a helpful simplification. It enables us to draw a graph like Fig. 5.7. A third justification for the one-dimensional treatment is that facts about the speed of sound, the concept of wavelength, and the formula $v = f\lambda$ are the same whether the number of spatial dimensions is one, two, or three.

Fig. 5.8 There is a source of
sound at the origin where the
x and y axes meet. The *solid
circles* indicate the peaks in
air pressure as seen in a
snapshot taken at time t_1.
Successive peaks are
separated by one wavelength,
as expected. Slightly later, at
time t_2, the wave has moved
outward, and the peaks are
now at the positions indicated
by the *dashed circles*

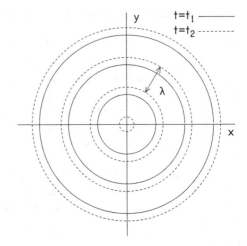

Two-Dimensions Next, we ought to think about more spatial dimensions. Suppose
there are two spatial dimensions. Then the sound waves are analogous to the ripples
of water waves on the surface of a pool. A surface has two spatial dimensions. We
can show a wave traveling on the surface by plotting only the *peaks* of the waveform
at two different instants in time, time t_1 and time t_2 a moment later. In the interval
between these two times the peaks of the waveform will move, as shown in Fig. 5.8.

Three-Dimensions From two spatial dimensions it is not hard to make the jump
to three. Look at Fig. 5.8 and imagine that the concentric circles are turned into
concentric hollow shells. As time goes on the shells expand like the surface of a
spherical balloon as it is blown up. The shells expand at the speed of sound. That
gives you the image of how the peaks of a sound wave move in three dimensions.

5.5   Unit Analysis

The term "Unit analysis" (also known as "Dimensional analysis") refers to
the fact that equations must make sense in terms of their physical units. This
logical fact can help you avoid errors in applying equations. It can help you
remember an equation that you have forgotten. It can even enable you to make
a good guess about an equation that you never learned!

The best way to learn about unit analysis is with a few examples.

You know that the distance you travel (d) is equal to your speed (v) multiplied
by the duration of your travel (t), or $d = vt$. In terms of units, this might
be stated as, "(miles) equals (miles per hour) multiplied by (hours)." As an
equation this is:

$$\text{miles} = \frac{\text{miles}}{\text{hour}} \times \text{hours}.$$

We have used the fact that the word "per" means "divided by." The key point is that the units of hours cancel in the numerator and denominator, leading to the logical conclusion that miles equal miles. Unit analysis enforces that kind of logic.

Notice that unit analysis can prevent you from making an error such as multiplying a speed in miles per hour by the number of *minutes* you travel.

Unit analysis applies to every equation we study. For instance, Eq. (5.2) says $v = f\lambda$. In terms of units

$$\text{meters per second} = \frac{\text{cycles}}{\text{second}} \times \frac{\text{meters}}{\text{cycle}}.$$

The units of cycles cancel, and this equation reduces to a logical result:

$$\text{meters per second} = \frac{\text{meters}}{\text{second}}.$$

Notice that unit analysis helps you avoid an error in writing the equation. For instance, $v = \lambda/f$ is a nonsense formula. It says that a speed (v) is equal to a length (λ) multiplied by a time ($1/f$), but you know that a speed is equal to a length *divided* by a time.

As another example, Chap. 2 says that frequency, f, has units of inverse time. Therefore, $1/f$ has units of seconds. Formally we say that period, T, is given by $T = 1/f$, or,

$$\text{seconds per cycle} = \frac{1}{\text{cycles/second}}.$$

You can improve your life by applying unit analysis to every practical calculation you do.

Exercises

Exercise 1, Constant pressure

The pressure of the sound shown in Fig. 5.2 is more than 101,000 Pa. Bozo says that such a big pressure is bound to be painful. Set Bozo straight on the role of this constant part of the pressure.

Exercise 2, Check those powers of ten

The first page of this chapter says that 100,000 N/m^2 is equivalent to 10 N/cm^2. Is that really right? Consider that 1 m is 100 cm. Why isn't 100,000 N/m^2 equivalent to 1,000 N/cm^2?

Exercise 3, Pressure conversion

Your goal is to show that 14.7 pounds per square inch is equivalent to about 101,000 N/m^2. You begin by realizing that pressure is force per unit area. Therefore, you need English-metric conversions for force and for area. For force: 4.45 N/pound. For distance: 0.0254 m/in., therefore, 0.000645 m^2/in.^{2}. Can you work it from here?

Exercise 4, Extreme weather

On a really cold day the temperature is −18 °C (0 °F). On a really hot day it is +40 (104 °F). What is the speed of sound at −18? At +40?

Exercise 5, Americanize the equation

Our formula Eq. (5.1) for the speed of sound as a function of temperature assumes that temperature is given on the Celsius (also known as Centigrade) scale. On this scale water freezes at zero and water boils at 100°. Most Americans are more familiar with the Fahrenheit scale where water freezes at 32° and boils at 212°. A temperature on the Fahrenheit scale T_F can be converted to a temperature on the Celsius scale T_C with the formula

$$T_C = \frac{5}{9}(T_F - 32). \tag{5.6}$$

(a) Show that the speed of sound, in meters per second, is given by

$$v = 321 + \frac{1}{3}T_F. \tag{5.7}$$

(b) Room temperature on the Fahrenheit scale is about 69°. Show that formula (5.7) leads to a speed of sound of 344 m/s at this temperature as you would expect.
(c) To further Americanize the equation, show that the speed of sound in feet per second is given by

$$v = 1053 + 1.1\,T_F. \tag{5.8}$$

Use the fact that 1 ft is exactly 0.3048 m.

Exercise 6, A fair start

(a) How long does it take for sound to travel 10 m? (b) The runner in the far lane is 10 m farther away from the starter than the runner in the near lane. When the starter fires the starting pistol, the near runner hears it first. Could this make a significant difference?

Exercise 7, Stormy weather

Thunder and lightning are produced simultaneously. If you see the lighting 2 s before you hear the thunder, how far away did lightning strike?

Exercise 8, Mach 1

The text says that the speed of sound at room temperature is 1,128 feet per second or 769 miles per hour. Use the fact that there are 5,280 ft in a mile to show that this equivalence is correct.

Exercise 9, Sound vs light
The section called "Sound vs light" gives the range of audible sound frequencies and the range of visible light frequencies. Which range is broader?

Exercise 10, Longest and shortest
Nominally, the limits of human hearing are 20 and 20,000 Hz. Find the wavelengths of those two waves in air. Here, and in exercises below, assume that the air is at room temperature.

Exercise 11, Medium long
What is the frequency of a tone with a wavelength of 1 m?

Exercise 12, Frequency and wavelength
Equations (5.2), (5.3), and (5.5) are all variations on the same equation. Is one more fundamental than the others? Knowing what you do about waves that travel from one medium into another, can you make the case that Eq. (5.5) $[\lambda = v/f]$ is the most fundamental?

Exercise 13, Making waves
Clap your hands four times with a perfectly regular rhythm so that there is 1/2 s between each clap. Think about the hand claps as a periodic signal. It is not a sine wave signal, but it is periodic—at least for a few seconds. Now look at Fig. 5.8. Could this apply to the hand clapping exercise? What is the period, the frequency, the wavelength?

Exercise 14, Unit analysis
By law, the maximum water usage of a new American toilet is 1.6 gpf. What do you think is meant by "gpf." Write the equation that describes how many gallons of water the toilet uses in a day. Your equation should include all the units, which is the main point of this exercise.

Exercise 15, Dispersion
The speed of a sound wave in open air does not depend on the frequency of the wave. We say that the propagation in air is *non-dispersive*. Some physical systems do have dispersion. For instance, when a light wave travels through a glass prism, the speed of light in the glass depends on the color (frequency) of the light. The dispersion of the prism separates the different colors that compose white light. Suppose that sound propagation were dispersive so that low-frequency sounds travel 10 % faster than high-frequency sounds. What would be the consequences for communication?

Exercise 16, Bending water?
Can transverse waves propagate in a fluid like water?

Exercise 17, Breaking waves
Breaking waves that you see coming up the beach are complicated. They are more complicated than the sound waves that we describe as longitudinal or transverse. (a) Make the case that the polarization of breaking waves is longitudinal. (b) Make the case that the polarization of breaking waves is transverse.

Exercise 18, Helium gas

Helium gas is much lighter than ordinary air, i.e. it is less dense. That is why it is used as a lifting gas for blimps. Because helium is so light, the speed of sound in helium is high, about 1000 m/s.

(a) Show that the speed of sound in helium is about three times faster than the speed of sound in air.

(b) Show that the wavelength of a 1000-Hz tone in helium gas is about 1 meter.

(c) Explain why the frequency of a 1000-Hz tone in helium gas is still 1000 Hz.

♠

Chapter 6
Wave Properties

This chapter describes some properties of waves—properties like interference, beats, reflection, and refraction. Naturally, the focus will be on sound waves, but the principles apply equally to light waves. Because of the unifying principles of wave physics, you get two for one—once you've learned the acoustics, you automatically know the optics.

6.1 Wave Addition

Your date is talking to you across the table, and the band is playing in the background. You hear them both. You may imagine that there are two sound waves, one from your date and the other from the band and that both sound waves are propagated to your ears. That description is not entirely wrong, but a much better description is that the waves from the two sources are added in the air, and your ear is exposed to only a single wave. The single wave is the sum of the waves from the two sources. The addition of waves is the addition of pressure waves. Because pressure waves can be positive or negative there are possibilities for reinforcement and cancellation. Reinforcement occurs when two waves are added and they have the same sign—both are positive or both are negative. Then the summed wave is more positive or more negative than either of its constituents. Cancellation occurs when the constituent waves have different signs. If one wave leads to a positive pressure at a particular place and time and the other wave leads to a negative pressure at that place and time, then the two waves tend to cancel each other when they both occur at the same time.

W.M. Hartmann, *Principles of Musical Acoustics*, Undergraduate Lecture Notes in Physics, DOI 10.1007/978-1-4614-6786-1_6,
© Springer Science+Business Media New York 2013

6.2   Interference

In the most general terms, interference is the same thing as the addition of several waves as described in the section above. Specifically though, when we talk about interference, we usually are thinking about a situation where the addition is particularly simple because the two sources are similar. Both sources emit the same waveform and have the same frequency. This simple situation is important because dramatic events can occur. As usual in science, simplicity leads to power.

If the waves from the two sources add up to reinforce one another we have *constructive interference*. If the waves from the two sources add up to cancel one another we have *destructive interference*. Constructive interference occurs at a point in space where the two waves have the same phase. When one is positive the other is positive. When the first one is negative the second one is also negative.

Destructive interference occurs when the two waves have opposite signs. When one is positive the other is negative. Two sine waves can be caused to have opposite signs by delaying one of them by half a period. Then the two sine waves become 180° out of phase. You will recall that what you get on the second half cycle of a sine wave is just the negative of what you got on the first half cycle. *Therefore, the key to completely destructive interference is a delay of half a period.*

Figure 6.1 shows two loudspeakers and a microphone. Speaker 1 is 1.06 m away from the microphone; speaker 2 is 1.4 m away from the microphone. Both speakers emit a sine tone with a frequency of 500 Hz so that the period is 2 ms.

This example happens to be a case of interference that is almost completely destructive. The reason is that the two waves arrive at the microphone almost 180° out of phase. That is because the extra time for the sound from speaker 2, compared to speaker 1, is half a period of the tone. Let's see why that is so. The time required for any waveform feature to get from speaker 1 to the microphone is t_1, $t_1 = D_1/v$, where v is the speed of sound, $v = 344$ m/s, as shown in the time-line of Fig. 6.2. The time for the feature to arrive from speaker 2 is t_2, $t_2 = D_2/v$. The extra time is $t_2 - t_1$, and it is easily calculated.

$$t_2 - t_1 = D_2/v - D_1/v = 1.4/344 - 1.06/344 \approx 0.001 \text{ seconds,} \qquad (6.1)$$

or 1 ms. Because the period of the tone is 2 ms, the extra time is half a period, just what is needed for cancellation.

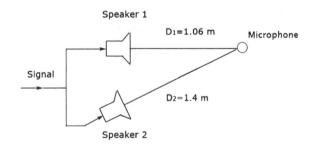

Fig. 6.1 Two loudspeakers and a microphone. The same sine signal is sent to both speakers

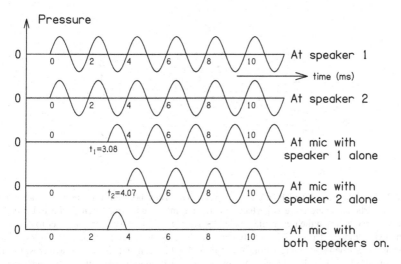

Fig. 6.2 Delay $t_1 = 1.06/344$. Delay $t_2 = 1.4/344$. The two waves cancel each other at the microphone position

Note that if the extra travel time were a full period, then the waves from the two speakers would be in phase, and there would be completely constructive interference.

Note that if the extra travel time were one-and-a-half periods, then the waves from the two speakers would be 180° out of phase again and the interference would be destructive.

It should be clear that the condition for destructive interference is that the extra travel time should be an odd number of half periods, i.e., $T/2, 3T/2, 5T/2, 7T/2$, etc. The formula for destructive interference is then

$$t_2 - t_1 = (2n + 1)T/2, \tag{6.2}$$

where n is an integer. [Note that $(2n + 1)$ is just a way to describe any odd integer. Try it.]

We can express that result in terms of the extra distance that the wave from speaker 2 must go. Call that extra distance $D_2 - D_1$. Then destructive interference occurs when the extra distance is an odd number of half wavelengths, i.e.

$$D_2 - D_1 = (2n + 1)\lambda/2. \tag{6.3}$$

To prove that Eq. (6.3) is correct, we can start with Eq. (6.2) for cancellation described in terms of delay, then multiply both sides of the equation by the speed of sound to get distances:

$$D_2 - D_1 = v(t_2 - t_1) = (2n + 1)vT/2 = (2n + 1)\lambda/2. \tag{6.4}$$

The proof has used the fact that $vT = \lambda$.

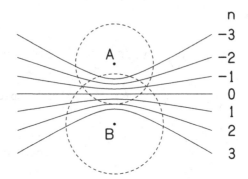

Fig. 6.3 Two sources, A and B, are separated by 3 m. They both emit a tone with wavelength 0.5 m. Hyperbolas where there is perfect constructive interference are labeled with index n. For instance, every point on the hyperbola labeled "−1" is exactly one wavelength closer to A than to B. The hyperbola labeled "−2" is exactly two wavelengths closer, etc. The *dashed circles* show a snapshot taken when a wave crest from A has moved out by 2 m and an earlier wave crest from B has moved out by 2.5 m. The difference is exactly one wavelength and so the circles intersect on the line labeled "−1"

Interference, as described above, is a spatial effect. Two waves having exactly the same frequency may add constructively or destructively at different points in space. At some points in space they may cancel almost entirely so that a dead spot occurs for the particular frequency in question, and the spot remains dead for all time. Interference, as described above, uses spatial differences to create the particular phase relationships between the two waves so that constructive or destructive interference occurs. For instance, a path length difference of 0.34 m led to a 180-degree phase difference in the example.

A two-dimensional view of constructive interference between two sources, A and B, can be seen in Fig. 6.3. Constructive interference occurs at all points where the distance from source A and the distance from source B differ by one wavelength, or two wavelengths, or three, etc., i.e.

$$D_A - D_B = n\lambda \tag{6.5}$$

For a particular value of n we are looking for all the points where the difference between the distance from point A (D_A) and the distance from point B (D_B) differ by a certain amount ($n\lambda$). It is a fact of geometry that all points that fulfil a requirement like that in two dimensions lie on a hyperbola. Such hyperbolas are shown by the solid lines in the figure. At every point on those lines there is perfect constructive interference.

6.3 Beats

The phenomenon called "beats" is also a form of interference. There are maxima and minima from constructive and destructive addition of tones, but beats take place in time and not in space. At some times there is constructive interference and at

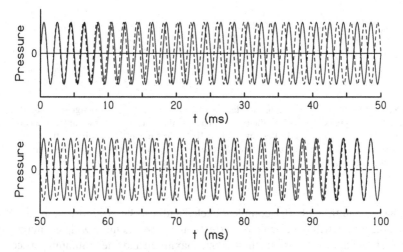

Fig. 6.4 Two waves, 500 and 510 Hz. The time axis begins at the *upper left* and continues on the second graph below. Beats occur when the solid wave and the dashed wave are added together. The pattern shown here repeats for as long as the two tones persist

other times there is destructive interference. When the interference is constructive it is constructive at all points in space, and when it is destructive it is destructive at all points in space.

Beats occur between two sine tones that have *almost* the same frequency but *not exactly* the same frequency. The beat rate (the number of beats per second) is equal to the difference between the two frequencies. If the two frequencies are f_1 and f_2, then the beat rate is

$$f_{beat} = f_2 - f_1 \qquad (6.6)$$

For instance, suppose you strike two tuning forks, and one has a frequency of 500 Hz while the other has a frequency of 501 Hz. The difference is $501 - 500 = 1$, or one beat per second. That means that there will be one maximum, when the sum is loud, and one minimum, when the sum is quiet, every second. It is not hard to see why this is so.

Suppose that we measure the two waves from the two tuning forks when each is at a positive maximum. They are "in phase." Then the sum will be as loud as it can be. As time progresses the two waves will become out of phase because the forks have different frequencies. After half a second the 500-Hz wave will have executed 250 complete cycles and it will be at a positive maximum again. However, the 251-Hz wave will have executed 250.5 cycles, and the extra half a cycle will mean that this wave is at a negative maximum. The positive maximum and negative maximum will cancel and the sum will be quiet. After a full second, the 500-Hz wave will have gone through 500 cycles and be at a positive peak; the 501-Hz wave

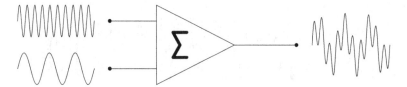

Fig. 6.5 A two-input mixer. As usual, signals go from left to right. Two signals come into the mixer and one signal comes out on the right. The signal that comes out is the sum of the two input signals. The *sigma symbol* indicates summation. To create a musical recording or the sound track of a film, audio engineers use mixers that add dozens of signals together. A 48-input mixer is not uncommon. The principles of interference among audio signals still apply

will have gone through 501 complete cycles and also be at a positive peak. As a result, the waves will once again be perfectly in phase and there will be a maximum again. This example shows that there is one maximum and one minimum per second when the frequency difference is 1 Hz.

Figure 6.4 shows what happens when the frequencies are 500 and 510 Hz. Now there are ten beats per second. Ten beats per second is rather fast but one can still hear them as a rapidly varying loudness. Again, we choose to start the clock when both waves are together. After 50 ms (0.05 s) the 500-Hz wave will have gone through 25 cycles but the 510-Hz wave will have gone through 25.5 cycles. The waves will be out of phase and will cancel. After 100 ms the two waves will have gone through 50 and 51 complete cycles and will be in phase again.

6.4   Audio Analogies

The previous sections have described interference and beats in acoustical terms. The outputs of two loudspeakers were added in the air to make a single wave that appeared at the microphone. The outputs of two tuning forks were added in the air to create a single beating wave at a listener's ear.

Because we often deal with signals in electronic (audio) form, it is important to note that interference and beats also occur between *audio* signals that are added together. The electronic device that adds audio signals together is called a "mixer." Figure 6.5 shows a mixer with two inputs. If these two inputs contain sine tones that happen to have the same frequency and the same amplitude but are 180° out of phase, then the sine tone in the output will cancel just as surely as it did in the two-loudspeaker experiment. Alternatively if the two inputs contain sine tones with frequencies of 500 and 510 Hz, then the output will contain a sine tone that beats ten times per second. One could see that beating tone by connecting the output to an oscilloscope.

6.5 Generalization

The concepts of interference and beats have been presented above using sine tones as examples. It was natural to use sine tones because the interference and beating effects are determined by frequency and a sine tone has only a single frequency. That makes it simple. It is important to realize that these effects also take place for the more complicated signals of speech and music because sine tones are present as components in complicated signals. An interference effect or a beating effect that occurs for a sine tone in isolation also occurs in the same way when that sine tone is a component of a complex tone.

Telephone Dial Tone To illustrate the generalization above about complex signals, consider the dial tone in an American telephone. The dial tone is a sum of two sine waves, 350 and 440 Hz. This sum makes a complex wave that sounds like a musical chord. Suppose you did not know what the frequencies were and you wanted to find out. An oscilloscope tracing would be a mess because the signal is not particularly periodic. For the same reason, a frequency counter would do you no good either.

What you *could* do is to add another sine tone with adjustable frequency and listen for the beats. You can choose to get beats from either the 350-Hz component or the 440-Hz component. When you adjust the frequency to 345 Hz you hear 5 beats per second because of the 350-Hz component. When you adjust the frequency to 443 Hz you hear 3 beats per second because of the 440-Hz component. By counting the beats for different frequencies of the adjustable tone near 350 and 440 Hz you could learn the exact frequencies of the two components in the dial tone.

Electronic Inversion The present chapter opened with a discussion of wave addition and interference. It showed that a 500-Hz tone can be cancelled by adding another 500-Hz tone if one of these tones is delayed by 1 ms compared to the other. That is because 1 ms is one-half of the period—just what is needed to make the two tones 180° out of phase. Similarly, a 400-Hz tone can be cancelled by adding another 400-Hz tone that has been delayed by 1.25 ms. This kind of cancellation works perfectly for sine tones.

But what happens for a complex tone? Suppose you want to place a single speaker so as to cancel a complex tone that is the sum of a 400-Hz tone and a 500-Hz tone, using a set up like Fig. 6.1. You could use a delay of 1.25 ms to cancel the 400-Hz part of the tone, or you could use a delay of 1 ms to cancel the 500-Hz part of the tone, but there is no delay that would cancel both components of the tone.

Within the domain of acoustics there is no solution to this complex cancellation problem. But electronically there is a solution because electronics allows for waveform inversion by a device called (appropriately) an "inverter." An inverter has an input and an output. Whenever the input voltage is positive, the output voltage is negative—and with the same magnitude. Thus, at an instant of time when the input is 3.45 V, the output is −3.45 V. Similarly, when the input is negative, the output is positive, again with the same magnitude. Therefore, the inverter happens to delay

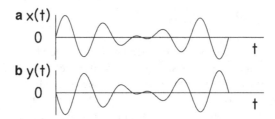

Fig. 6.6 (a) Waveform $x(t)$ is the sum of a 400-Hz sine and a 500-Hz sine. (b) Waveform $y(t)$ is the result of passing waveform x through an inverter. At any instant of time y is the negative of x. If x and y are added together in a mixer the output of the mixer is zero because x and y cancel completely

each frequency component by exactly the right amount to turn the waveform upside down. The inverted waveform $y(t)$ could be used to cancel the input waveform $x(t)$, as described in Fig. 6.6.

Noise-Cancelling Headphones

Our high-speed industrialized world is a noisy place. The sounds of machinery and transportation can mask other sounds that we need to hear or want to hear. Noise can be more than just annoying. It can be a hazard to health and safety. Wouldn't it be great if we could use electronics and the concept of interference to cancel the noise around us? The idea is simple in principle. All we need to do is to pick up the annoying sound with a microphone to convert it into an audio (electronic) signal, then invert the signal, and finally play it back through a loudspeaker. The signal from the loudspeaker should then cancel the annoying sound. This is the concept of *active noise cancellation*.

In fact, it is almost impossible with current technology to achieve active noise cancellation over a three-dimensional region of space, except for the very longest wavelengths (lowest frequencies). However, there are several acoustical circumstances where active noise cancellation can be successful. One application is in ducts where the sound is confined, and there is only one important direction of propagation, namely along the length of the duct.

The most common application of active noise cancellation is in headphones. Headphones are another confined environment, and they all contain little drivers with diaphragms that reproduce speech and music. In noise-cancelling headphones, the phone for each ear contains a tiny microphone to pick up the outside noise. The signal from the microphone is inverted and sent back, to be mixed with the desired speech or music. The combined signal is then sent to the headphone diaphragm driver. The result is that the inverted noise cancels the noise from outside and the user can hear the speech and music better. Alternatively, the user can disconnect the source of speech or music and simply listen to quiet.

Noise-cancelling headphones work well for low-frequency sounds where the wavelengths are long compared to the size of the headphone ear cup. Commercially available units are effective below 500 Hz but not at higher frequencies. Because much of the noise in aircraft and other vehicles is actually below 500 Hz, noise-cancelling headphones are useful. However, the sounds of intelligible human speech extend well above 500 Hz. Thus, noise-cancelling headphones can eliminate the low rumble inside an airplane, leaving a high-frequency rushing sound, but they do not prevent you from being disturbed by people using their cell phones while you are waiting in the airport.

6.6 Reflection

Waves can be reflected. A wave is reflected when it abruptly changes its direction of propagation. Suppose that you are in a room at night with a single lamp burning. If you look directly at the lamp you see a source of direct light. If you look at anything else in the room you are seeing reflected light. Similarly, sound waves are reflected from surfaces in a room. A wave is reflected when it encounters a change in medium for propagation. Both light and sound are reflected from a wall because the material of the wall is different from ordinary air.

There are two kinds of reflection depending on the regularity of the reflected wave. A mirror leads to *specular* reflection. You can see your face in a mirror. A typical wall leads to *diffuse* reflection. You cannot normally see your face in a wall. The difference can be seen in Fig. 6.7.

The specular reflection preserves the order of the rays and therefore preserves the image on reflection like a mirror. The diffuse reflection scrambles the order.

Whether a reflection is specular or diffuse depends on the size of the bumps on the surface compared to the wavelength. If the bumps are small compared to

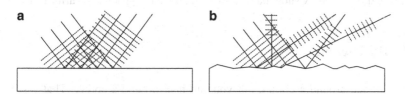

Fig. 6.7 (a) A wave comes in from the upper left. Four long, parallel lines show the direction of propagation of the rays, and the spaces between the wavefronts (shown *dashed*) indicate the wavelength. The wave is specularly reflected because the reflecting surface has irregularities that are much smaller than the wavelength. (b) The same wave comes in from the upper left, but it is diffusely reflected because the bumps on the surface are comparable to the wavelength or larger than the wavelength

Warm

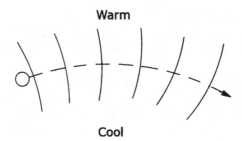

Cool

Fig. 6.8 The figure shows six *lines* indicating the pressure peaks in a traveling sound wave. The *lines* are separated by a wavelength. The wavelength is longer in warm air where the speed of sound is greater. The direction of propagation is perpendicular to the *lines* that show the peaks. Therefore, the direction of propagation is bent downwards toward the cool region

the wavelength the reflection is specular. If the bumps are large compared to the wavelength the reflection is diffuse. A specular reflection of sound waves in a room can lead to focusing of sound—often leading to an uneven sound distribution in the room.

6.7 Refraction

Refraction of a wave is different from reflection because the change of direction is gradual and not abrupt. A lens refracts a light wave. For instance, eyeglasses consist of lenses that make slight changes in the direction of the rays of light in order to correct for defects in the optics of the user's eyes. Lenses work because the conditions for propagation of light in glass are slightly different from the conditions in air or vacuum.

Sound waves can also be refracted when the conditions for propagation change. For instance, when sound travels from a region of cool air to a region of warm air, the wavefront moves faster in the warm air and bends toward the direction of the cool air. This effect occurs on a sunny winter day when the air immediately above the ground is cooled by snow but the air well above the ground is warm (Fig. 6.8).

6.8 Diffraction

You can't see around a corner, but you can hear around a corner. That essential difference between vision and hearing is the result of the very different wavelengths of light and sound. While visible light has a wavelength of about 500 nm (500×10^{-9} m, or 5×10^{-7} m), audible sound has a wavelength as long as 7 m, about ten million times longer than light. Sound waves bend around corners. Light waves bend around corners too—just ten million times less.

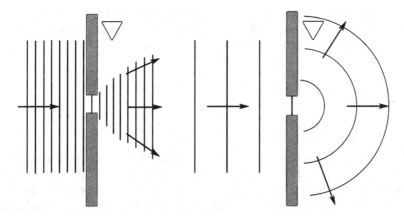

Fig. 6.9 A sound wave travels toward a door opening. In the figure on the *left* the sound wavelength is smaller than the door opening. In the figure on the *right* the wavelength is larger than the door opening. The large-wavelength sound spreads out beyond the door opening because of diffraction. Thus, a listener located at the position of the triangle can hear the large-wavelength sound

The ability of a wave to bend around a solid object is known as diffraction. The amount that a wave can bend depends on the ratio of the wavelength to the length of the bend. Diffraction also is involved when a sound wave passes through an opening such as a door. Then the amount of bending depends on the ratio of the wavelength of the sound to the width of the door opening. If the wavelength is small compared to the opening, then the wave goes straight through and does not bend much. Thus short-wavelength (high-frequency) sound behaves somewhat like light would behave. By contrast, long-wavelength (low-frequency) sound bends around the door opening, as shown in Fig. 6.9.

6.9 Segregation

We now return briefly to the example that began this chapter—your date and the band. This example illustrates the truly remarkable human capacity to selectively attend to individual sound sources. The sounds you want to hear are normally acoustically mixed in the air together with many other sounds. Moreover, the sounds that come directly to your ears are mixed with reflections from the surfaces in a room, creating a still more complicated sound field. As a listener, you sample this sound field with your two ears, and from this mixed up mess you extract the images of individual sources, e.g., your date and the band. This human ability is called "source segregation." It is a very impressive bit of signal processing. Scientists who study this aspect of human hearing for a living have yet to create a computer program that can segregate sound sources as well as the human brain. The source segregation problem is one of the greatest challenges in the field of artificial intelligence, but you routinely solve the problem without even thinking.

Exercises

Exercise 1, Magic wavelengths, magic frequencies
Two loudspeakers are connected so that they radiate the same sine tone. You are 3 m away from one loudspeaker and 3.4 m away from the other. (a) Find two values of the wavelength for which cancellation occurs. (b) What are the two frequencies for which cancellation occurs? (c) Are there more than two different wavelengths for which cancellation occurs?

Exercise 2, Total destruction
You are 1 m away from a source of a 1,000-Hz sine tone. (a) Where would you put a second source, driven by the same 1,000-Hz sine signal, in order to get complete cancellation? (b) Does it matter what the amplitudes of the two tones are?

Exercise 3, Stereo
In a standard stereo audio system, there are two loudspeakers. The loudspeakers can be connected wrongly so that they are out of phase. Why is this error more important for low frequencies than for high?

Exercise 4, Beats
You hear two tones with a beat rate of 5 Hz. One tone has a frequency of 440 Hz. What are possible frequencies for the other tone?

Exercise 5, The dial tone experiment
Fig. 6.10 depicts the dial tone experiment. The added tone at 345 Hz has an amplitude that is somewhat smaller than the amplitude of the dial-tone component at 350 Hz. Explain why the beats would be stronger if the added tone had the same amplitude as the component at 350 Hz.

Fig. 6.10 The *solid lines* show the amplitude spectrum of the telephone dial tone with components at 350 and 440 Hz. The *dashed line* shows a sine tone that can be added at 345 Hz to cause beats, five per second, to expose the 350 Hz tone

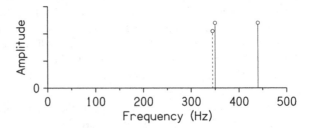

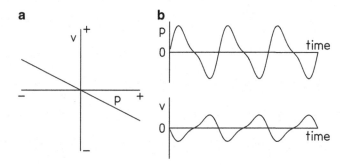

Fig. 6.11 Figure for Exercise 10. The input/output relationship of a linear transducer is a straight line. How would you describe this line?

Exercise 6, How smooth is smooth?

How smooth does a mirror need to be? Answer this question by working out the wavelength of light. Remember that the speed of light is 3×10^8 m/s. Recall that frequency of visible light is about 6×10^{14} Hz.

Exercise 7, How rough is rough?

Imagine a rough stone wall where some stones randomly protrude a few inches. How does a 100-Hz sound wave reflect from this wall? How about a 10,000-Hz wave? The answers to this exercise illustrate the general principle that a surface appears to be rough or smooth depending on the wavelength of the radiation that is used to examine the surface.

Exercise 8, Sound channels by refraction

With a complicated temperature profile, refraction can lead to narrow channels that trap sound waves. Which profile leads to a channel: (1) a layer of warm air trapped between layers of cool air, or (2) a layer of cool air trapped between layers of warm air?

Exercise 9, Elective deafness

Although it seems like blocking out the loud sounds of our busy and noisy world would be a good idea, having perfect noise-cancelling headphones would have its disadvantages. Explain why you would not want to completely knock out the noise of everyday life.

Exercise 10, Another linear transducer

Figure 4.2 shows the input/output relationship for a linear transducer that converts pressure waves into electrical waves. Figure 6.11 is similar but different. Explain how the difference in parts (a) of the figures leads to the difference in the voltage *v* in parts (b).

♠

Chapter 7
Standing Waves

The two previous chapters have described wave motion with particular reference to *traveling* waves. Traveling waves carry energy and information from one point in space to another. This chapter begins a study of *standing* waves. Standing waves turn out to be the modes of vibration of important classes of musical instruments. Accordingly, the chapter quickly moves on to the vibration of guitar strings.

7.1 Standing Waves in General

A traveling wave moves; a standing wave does not. But that does not mean that a standing wave does not change with time. It just changes differently from a traveling wave. The difference between a traveling wave and a standing wave can be seen in a series of snapshots.

Figure 7.1 is a snapshot of a 500-Hz wave. The period of the wave is 2 ms (1,000/500). The wavelength is $344/500 = 0.69$ m, and you can see that wavelength in Fig. 7.1 because the horizontal axis is a spatial axis, x.

If this wave is a *traveling* wave, then it moves rigidly in time. If the wave is traveling from left to right, and we look at it 0.25 ms later (one-eighth of a period), then we find that the wave has moved to the position shown by the dashed line in Fig. 7.2.

If this wave is a *standing* wave and we look at it 0.25 ms later (one-eighth of a period), the wave looks like the dashed curve in Fig. 7.3. It has not moved. It has just changed its size.

We continue to track the standing wave. If we look at the standing wave after another 0.25 ms (total time of $t_3 = 0.5$ ms, or one quarter of a period), we find that the standing wave is momentarily totally flat, as shown in Fig. 7.4.

After another 0.25 ms (total time of $t_4 = 0.75$ ms or three-eighths of a period) we find that the 500-Hz standing wave has come out the other side. It has changed from the original solid curve to the dashed curve in Fig. 7.5.

W.M. Hartmann, *Principles of Musical Acoustics*, Undergraduate Lecture Notes
in Physics, DOI 10.1007/978-1-4614-6786-1_7,
© Springer Science+Business Media New York 2013

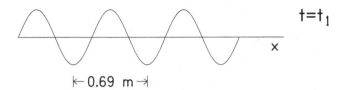

$t=t_1$

$\leftarrow 0.69\ m \rightarrow$

Fig. 7.1 A snapshot of a wave, taken at time t_1. This snapshot serves as the reference for snapshots to follow

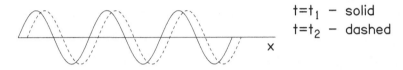

$t=t_1$ — solid
$t=t_2$ — dashed

Fig. 7.2 As time goes on, the 500-Hz traveling wave moves from left to right as time increases from t_1 to t_2

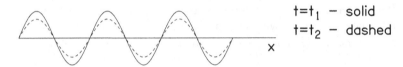

$t=t_1$ — solid
$t=t_2$ — dashed

Fig. 7.3 In contrast to the traveling wave, the standing wave does not move

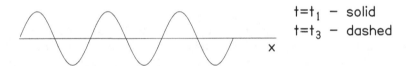

$t=t_1$ — solid
$t=t_3$ — dashed

Fig. 7.4 The snapshot of the standing wave taken at time t_3 is shown by a *dashed line*, but that line is hidden by the x axis

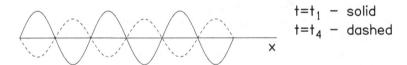

$t=t_1$ — solid
$t=t_4$ — dashed

Fig. 7.5 The snapshot of the standing wave taken at time t_4 is shown by a *dashed line*

If we look at the 500-Hz standing wave after another 0.25 ms (total time of $t_5 =$ 1 ms, or one-half of a period), we find that it has changed from the original solid curve to the dashed curve in Fig. 7.6.

Figure 7.6 shows how standing waves are normally represented, by two snapshots, one showing a maximum amplitude and the other, taken half a period later, showing a maximum in the other direction.

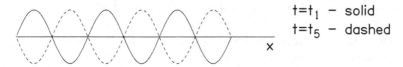

Fig. 7.6 The snapshot of the standing wave taken at time t_5 is shown by a *dashed line*. The difference in times, $t_5 - t_1$ equals half a period, $T/2$

As time goes on, the standing wave continues to evolve. The graphs of the continued motion look just like the graphs 7.2 through 7.6 played in reverse. For instance, after 1.25 ms the 500-Hz wave looks like it did at time t_4, and at 1.5 ms the wave looks like it did at time t_3. After a full period, the wave looks like it did at the beginning at time t_1.

A clear difference between a traveling wave and a standing wave is that with a standing wave there are places in space (values of location x) where there is never any displacement. No matter how long we wait at one of these places, there is never any action. Such places are called "nodes." By contrast, a traveling wave ultimately causes every point in space to experience some displacement. If there is no action now at your spot, be patient—the crest of a traveling wave will come by in a millisecond or so.

7.2 Standing Waves on a String

Standing waves don't just happen for no good reason. If there is a disturbance in a medium, such as a disturbance in the air or on a string, then the disturbance will travel as a traveling wave. A standing wave is caused by reflections in the system. For instance, if the string is stretched between two boundaries, then there will be reflections of the displacement wave at the boundaries. The key to understanding the standing wave are the conditions at the boundaries. Mathematicians call these "boundary conditions."

A guitar string is stretched between the bridge and the nut. A wave on the string is reflected from both these ends. The boundaries at the bridge and nut are places where the string is not allowed to vibrate. That is the key to the standing waves on the guitar string (or any other stretched string). The boundary conditions insist that there are no displacements at the ends of the string. The standing waves that are allowed to exist must be waves that satisfy those boundary conditions. That means that the wavelengths of the allowed vibrations must somehow "fit" onto the string.

The wave below has a wavelength that works. The wavelength is twice the length of the string so that half a wave fits perfectly onto the string. If the string is $L = 65$ cm long, then the wavelength is $\lambda = 130$ cm or $\lambda = 1.3$ m. This is the longest wavelength that will fit on the string. This pattern of vibration of the string is the *first* mode. Because this is the mode with the longest wavelength, this mode has the lowest frequency. The frequency depends on the speed of sound on the string (Fig. 7.7).

Suppose that the speed of sound on the string is $v_s = 107$ m/s. Then the frequency of this mode of vibration is given by

$$f = v_s/\lambda = 107/1.3 = 82.3 \text{ Hz.} \tag{7.1}$$

This is the fundamental frequency of the lowest string on a guitar. The note is called "E."

Because the wavelength is twice the length of the string, we can write the formula as

$$f = f_1 = v_s/(2L) \tag{7.2}$$

The second mode of vibration of the string has a complete wavelength fitting into the length of the string. A better way to say it is that two half-wavelengths will fit (Fig. 7.8).

The frequency of the second mode of vibration of the string is then

$$f_2 = v_s/(L) = v_s/(2L) \times 2 = 107/1.3 \times 2 = 165 \text{ Hz.} \tag{7.3}$$

The third mode of vibration has three half-wavelengths fitting into the length of the string (Fig. 7.9).

The frequency of the third mode is

$$f_3 = v_s/\left(\frac{2}{3}L\right) = v_s/(2L) \times 3 = 107/(1.3) \times 3 = 247 \text{ Hz.} \tag{7.4}$$

Mode 1 has no nodes. Mode 2 has one node. Mode 3 has two nodes. If we were to continue, mode 4 would have three nodes, etc. (This counting of nodes does not

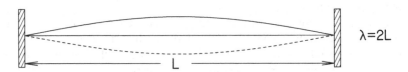

Fig. 7.7 The first mode of a stretched string gets half a wavelength into the length of the string

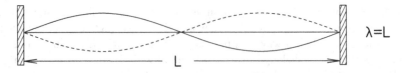

Fig. 7.8 The second mode of a stretched string gets a full wavelength into the length of the string

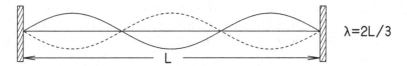

$\lambda = 2L/3$

Fig. 7.9 The third mode of a stretched string gets one and a half wavelengths (three halves) into the length of the string

count the two fixed ends of the string where there can never be any displacement.) It should be clear that there is no limit to the number of possible modes of vibration.

The different modes of vibration of the guitar string radiate by means of the guitar body. These different modes create the different *harmonics* of the guitar tone. The first mode creates the fundamental tone or first harmonic. The second mode of vibration creates the second harmonic, and so on. Each of the harmonic frequencies is an integer multiple of the fundamental frequency.

Harmonics The paragraph above used the word "harmonic." The word has been used before in this book, as in "simple harmonic motion." However, the paragraph above gives a second meaning to the word. Two or more components of a tone are said to be harmonics if their frequencies are small integer multiples of some base frequency. For instance, 150, 300, and 450 Hz are all integer multiples of 150 Hz. The integers are 1, 2, and 3, and the frequencies correspond to the first, second, and third harmonics.

For the stretched string, one can express the frequency of the nth mode (equivalently the nth harmonic) as

$$f_n = v_s/(2L) \times n \qquad (7.5)$$

This is a very simple relationship. As we study different musical instruments in the chapters that follow, the instruments will have physical modes of vibration that correspond to harmonics in the tone of the instrument, but the correspondence between modes and harmonics will not always be so simple.

7.3   The Guitar Player's Equation

To play melody and harmony, the guitar player must play notes with correct fundamental frequencies. The fundamental frequency of a note, or tone, is determined by the speed of sound and the length of the string. The speed of sound, in turn, depends on the tension in the string (F) and the linear mass density (μ) of the string according to the equation

$$v_s = \sqrt{\frac{F}{\mu}}. \qquad (7.6)$$

Here, F is the tension measured in Newtons. Recall that Newtons is a metric unit of force, equivalent to about 1/4 pound in the English system of units. Linear mass density μ is the mass per unit length of the string. It is measured in kilograms per meter (kg/m).

Putting the information from Eqs. (7.2) and (7.6) together gives the guitar player's equation for the fundamental frequency of a tone,

$$f_1 = \frac{1}{2L}\sqrt{\frac{F}{\mu}}. \tag{7.7}$$

Thus, three different string properties go into making the pitch of the guitar tone: length (L), tension (F), and linear density (μ).

7.4   The Stretched String: Some Observations

Modes: Really? The discussion of standing waves on a string made frequent use of the word "mode." This is not the first time you have seen this word. This word was introduced in Chap. 3 to describe a natural pattern of vibration of a physical system. Is it correct to use this word again to describe the standing wave? Absolutely—it is correct. The modes of vibration of a stretched string *are* standing waves. Recall that the requirements for a mode are that there be (1) a specific frequency, (2) a specific shape, and (3) an amplitude that can be determined by the user. You should be able to see how our definition of standing waves on the stretched string fulfills all those requirements.

The Musical Significance of the Stretched String: Harmonics Think about all the things there are in the world that you could bang on or blow on or otherwise cause to vibrate. Now think about the physical systems that are actually used in making music. Only small fraction of the possible things in the world make useful musical instruments, and the stretched string is one of them. The stretched string dominates our orchestras because of the violin family (includes violin, viola, cello, and bass viol) and it dominates our rock groups because of guitars. It is also the acoustical basis of the piano. This is really rather strange. The stretched string seems like an unlikely candidate for a musical instrument. It makes a very poor radiator. It has so little vibrating surface area that one always needs a sound board or amplification of some kind to make it at all useful musically.

There must be a reason why this one physical system, out of all the possibilities, is so dominant. The reason is that the modes of vibration have frequencies that are all integer multiples of a fundamental frequency. The second mode has a frequency that is twice the frequency of the first mode. The third mode has a frequency that is three times the frequency of the first mode, and so on. Because of this useful fact about the frequencies of the modes, the stretched string can create tones consisting of vibrating components that are harmonic. These components of a tone are sine waves and they are called "partials" because the entire tone is made up of many sine waves. Each sine component is only "part" of the entire wave. The partials of a tone made by a stretched string are harmonic because of the integer relationship between the frequencies of the partials. We can rewrite Eq. (7.5) to emphasize that fact,

$$f_n = n f_1. \tag{7.8}$$

This is an equation that is just about the frequencies of the modes, and ultimately about the frequencies of the partials. Most other physical systems have modes with frequencies that do not obey this special harmonic law.

Idealization and Reality When you pluck a guitar string your thumb pulls the string to the side and releases it. There is a restoring force that then causes the string to move toward equilibrium. The restoring force is responsible for the vibration of the string. Most of the restoring force is caused by the tension in the string. However, some of the restoring force comes from the *stiffness* of the string. To understand stiffness, imagine that the guitar string is not on the guitar but is lying loose on the table. There is no tension and yet if you try to kink the string tightly you encounter some resistance. The resistance is stiffness. This stiffness also contributes to the resistance to displacement when the string is stretched on the guitar in the usual way.

Our treatment of the stretched string, and the harmonic law concerning the mode frequencies, is based on the assumption that *all* of the force on the string comes from tension and none of it comes from stiffness. In other words, it is an idealization of the real string. Look at Eq. (7.7). There you see the importance of tension (F) but there is nothing about stiffness. When we discuss about string instruments later in this book, we will deal with real strings. As we will see at the time, what looks like a potential problem turns into a benefit.

Exercises

Exercise 1, High mode number
Draw the standing wave corresponding to the sixth mode of string vibration. How many nodes are there? (Don't count the two fixed ends of the string.)

Exercise 2, The A string
If the speed of sound on a string is 154 m/s, and the string length is 70 cm, what is the frequency of the first (or fundamental) mode of vibration? What is the frequency of the third harmonic?

Exercise 3, The guitar player's equation
Explain how the three parameters, μ, F, and L, are used in building, tuning, and playing a guitar.

Exercise 4, Standing and traveling
Explain how standing waves and traveling waves change with time. How are they different?

Exercise 5, How to play the guitar

To play different notes, a guitar player presses fingers down on the string to make it shorter. What is called an "open string" occurs when the player does not press down with a finger but uses the full string length (as the instrument was manufactured). The open E string has fundamental frequency of 82 Hz. Suppose it is 70 cm long. How short must the string be to play the note called "A" with a frequency of 110 Hz?

Exercise 6, A tale of two strings

The lowest fundamental frequency on a guitar is 82 Hz—the open E string. The top string on a guitar is also an E string. The open top string sounds the E two octaves higher. How much faster is the speed of sound on the top E string compared to the lowest string? Recall that going up by one octave corresponds to a doubling of the frequency.

Exercise 7, Bending notes

Bending an note means to change its frequency slightly in a slow and continuous way for musical effect. To play the blues you have to bend notes. How does a rock guitar player bend notes? What is the physical principle involved?

Exercise 8, The effect of tension

A guitar string has a fundamental frequency of 100 Hz. What does the frequency become if the tension in the string is doubled?

Exercise 9, More unit analysis

Chapter 5 introduced unit analysis with some simple examples. You might like to try unit analysis on a not-simple example such as Eq. (7.6) for the speed of sound on a string, v_s, given in meters per second.

$$v_s = \sqrt{\frac{F}{\mu}}.$$

The denominator is the linear mass density, and its units are kilograms per meter or (kg/m). The numerator is a force with units of Newtons, but that is not very helpful. The key to the units of force is Newton's second law, which says that force is equal to mass multiplied by an acceleration. Mass has units of kg, and acceleration has units of m/[(s)(s)]. Therefore, a Newton of force corresponds to kg m/[(s)(s)]. Show that the units on the right-hand side of this equation are m/s as expected. Don't forget to take the square root of the units. Note that a unit such as m/[(s)(s)] would normally be written as m/s^2.

Exercise 10, Role of the guitar body

From the *guitar player's equation*, Eq. (7.7), it looks as though the frequency of a guitar note, and hence its pitch is entirely determined by the vibrating string and is not related in any way to the body of the guitar. Can that possibly be right?

Fig. 7.10 Second mode of
vibration of a stretched string
at 19 different instants of time
separated by 1 ms

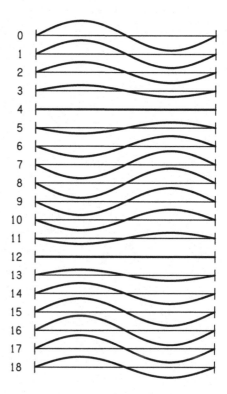

Exercise 11, Other musical instruments

Not all musical instruments use strings for vibration, however they still use standing waves to make sound. Two of the most common non-string instrument categories
are wind instruments and drums. What vibrates in these other instruments?

Exercise 12, One more guitar equation

Combine Eqs. (7.5) and (7.6) to obtain an equation for the frequency of the nth
harmonic of a guitar tone.

Exercise 13, Second mode through time

Figure 7.10 shows 19 consecutive snapshots of the second mode of vibration
taken at 1 ms intervals. Show that the frequency of this mode is 62.5 Hz. Show that
the frequency of the fundamental mode is 31.25 Hz.

♠

Chapter 8
Standing Waves in Pipes

8.1 Pipe with Both Ends Open

A cylindrical pipe (a flute, a marimba resonator, an automobile exhaust pipe) filled with ordinary air is a system that shows standing waves. If both ends of the pipe are open, the boundary conditions insist that the pressure at the open end is zero. Zero means that it is not overpressure (positive); it is not underpressure (negative); it is ordinary atmospheric pressure (zero).

A snapshot of the first mode of vibration looks like Fig. 8.1.

Here the solid curve is a snapshot taken at a time when air has rushed in from the two ends and piled up in the middle of the pipe creating a pressure maximum there. The dashed curve is a snapshot taken half a cycle later when the air has rushed out of the open ends and created a partial vacuum in the middle of the pipe.

A snapshot of the second mode of vibration looks like Fig. 8.2.

It should be clear that in this mode, the air is sloshing from one side of the pipe to the other, like the water in a bathtub. The solid line shows the pressure sloshed to the left side with a pressure valley on the right. Half a cycle later (dashed line) the pressure has sloshed to the right side.

The graphs of the modes of air in a cylindrical pipe are just the same as the graphs of the modes of a vibrating string. The physical systems are different, but the mathematical analysis is the same. If you know about the string there is almost nothing more to learn.

Just as in the case of the stretched string, the modes have wavelengths given by the equation $\lambda_n = 2L/n$ where n is an integer indicating the mode number. Therefore the frequencies are given by the formula

$$f_n = [v/(2L)]n \tag{8.1}$$

The only new thing to learn is that when dealing with the vibrations of air in a pipe, you must use the speed of sound in air for v. For the vibrations of a string you must use the speed of a transverse vibration wave on the string.

W.M. Hartmann, *Principles of Musical Acoustics*, Undergraduate Lecture Notes in Physics, DOI 10.1007/978-1-4614-6786-1_8,
© Springer Science+Business Media New York 2013

Pressure

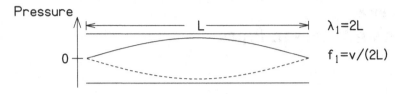

Fig. 8.1 Open–open pipe. The first mode has no nodes as we count them

Pressure

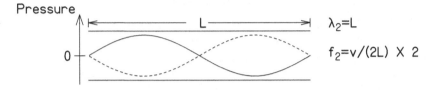

Fig. 8.2 Open–open pipe. The second mode has one node

Example:
If a cylindrical pipe is 90 cm long and open at both ends, then what is the frequency of the third mode of vibration?

Answer:

$$f_3 = v/(2L) \times 3 = 344/(2 \times 0.9) \times 3 = 573 \text{ Hz.} \tag{8.2}$$

8.2 Pipe with One End Open and One End Closed

Closing off one end of a pipe changes the physical system dramatically. The first mode in an open–closed pipe looks like Fig. 8.3. There is a pressure maximum at the closed end of a pipe as the air rushes in and jams against this closed end. Half a cycle later there is a pressure minimum at the closed end. Hence, the boundary condition for a standing wave at the closed end is an *antinode*. At the open end the boundary condition is zero pressure as usual.

The second mode looks like Fig. 8.4.

The third mode of vibration looks like Fig. 8.5.

There is an evident pattern. In the first mode one-quarter wavelength fits into the length of the pipe. In the second mode three-quarters of a wavelength fit. In the third mode five-quarters fit, and so on.

In connection with the frequencies f_1, f_2, and f_3, ..., of modes 1, 2 and 3, ..., symbol v again represents the speed of sound in air. The frequency f_1 for the open–closed pipe is $v/(4L)$ and this can be compared with the frequency for the open–open pipe, namely, $f_1 = v/(2L)$.

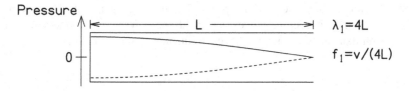

Fig. 8.3 Open–closed pipe. The first mode has only 1/4 wavelength in the pipe

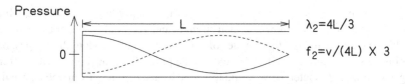

Fig. 8.4 Open–closed pipe. The second mode has a single node, 3/4 wavelength

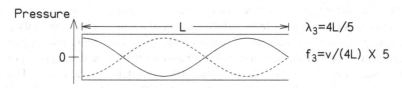

Fig. 8.5 Open–closed pipe. The third mode has 5/4 wavelengths in the pipe

Harmonics Again The modes of vibration of air in pipes have frequencies that are in a harmonic relationship. Therefore, used as musical instruments, pipes create tones with harmonic components. The pipe that is open at one end and closed at the other illustrates the warning given in Chap. 7 about more complicated relationships between modes and harmonics. Here, the first mode creates the first harmonic, the second mode creates the third harmonic, the third mode creates the fifth harmonic, and so on, all-odd numbered harmonics.

8.3 Playing a Pipe

If you blow across the open end of a pipe you can produce a tone. (Players of pan pipes are expert at this.) The tone occurs because blowing across the pipe excites the modes of vibration of the air in the pipe. More or less, all of the modes are excited at once, though it is usually the mode with the lowest frequency that is excited the most. Therefore, more or less all the modal frequencies are present in the tone at the same time. The *pitch* of the tone (what musicians call the "note") is normally given by the frequency of the lowest mode. This rule about pitch holds good both for pipes that are open at both ends and for pipes that are open at one end and closed at the other.

Shock exciting the modes of a pipe, by banging the open end of a pipe into your palm, also excites all the modes. Again, all the different mode frequencies are present simultaneously, though the lowest frequency modes dominate.

8.4   Thinking Critically

There is something rather strange about what this chapter has said about standing waves in pipes. It is actually worse than strange, and you ought to be questioning it.

This chapter related the frequencies of standing waves to the length of pipes, but nothing was ever said about the diameter of the pipes. Imagine a pipe that is 90 cm long and 1 km in diameter. How would this pipe be any different from open air in the great outdoors? One can't imagine that such a system would have standing waves like those shown in the figures above. And why doesn't the diameter appear in the equations for the frequencies of the standing waves? How does this make any sense?

These are good questions. In fact, the treatment of the modes in a pipe makes the assumption that the pipe diameter is small compared to the length. Such a pipe is called "thin." The fact that the diameter is small means that the pipe appears to be a one-dimensional system and the only dimension of interest is the length. That is why the length is the only dimension that appears in the equations.

If the diameter of the pipe is not much smaller than the length, then the picture described in this chapter does not apply. If the diameter is much smaller than the length, but not *very much* smaller, then the picture described in this chapter is a decent approximation to reality, and it can be improved by including open-end corrections, as described below.

8.5   Open-End Corrections

Our treatment of an open end of a pipe has not been exactly correct. We have assumed that the pressure goes to zero (equivalent to the outside atmospheric pressure) right at the open end of the pipe. In fact, the pressure wave extends a little bit beyond the end of the pipe. The point of zero pressure actually occurs slightly outside the pipe. The amount by which the pressure wave extends beyond the open end of the pipe depends on the pipe diameter. Therefore, for the purpose of calculating the resonant frequencies of a pipe with one or two open ends, the effective pipe length is longer than the measured pipe length. The measured pipe length should be increased by the end correction to find the effective pipe length. In equation form, if L is the measured length of the pipe and L_{end} is the end correction, then the effective length of the pipe is given by

$$L_{effective} = L + L_{end} \tag{8.3}$$

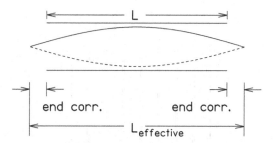

Fig. 8.6 Because the standing wave in a pipe extends beyond the ends of the pipe, the effective pipe length $L_{effective}$ is longer than the measured length L. The difference is the end correction. For a pipe like this one with two open ends, the end correction must be added twice to obtain the effective length

The end correction is equal to 0.61 times the radius of the pipe (0.305 times the diameter) for one open end. If both ends are open the end correction is twice as large (Fig. 8.6).

Example:
A pipe is 1 m long and open at both ends. Its diameter is 5 cm (0.05 m). What is the lowest resonant frequency of the pipe?

Answer:
The lowest resonant frequency is given by $v/(2L_{effective})$.
The effective length of the pipe is $L_{effective} = 1 + 0.05 \times 0.305 \times 2$ or 1.0305 m. Therefore, the lowest resonant frequency is

$$f = v/(2L_{effective}) = 344/(2 \cdot 1.0305) = 167 \text{ Hz.} \qquad (8.4)$$

Note that without the end correction we would have calculated a frequency of 172 Hz. $[v/(2L) = 344/2 = 172]$. The difference is significant.

Making a pan flute
The goal of this example is to construct a pan flute that plays the tones of a major scale starting on middle C and ending an octave higher. Such a scale has eight notes, called C, D, E, F, G, A, B and C, and so the pan flute will be made from eight tubes. They will be closed at one end.

From Appendix E we learn that middle C (called C_4) has a fundamental frequency of 261.6 Hz. To make the other notes, we choose an equal-tempered scale. Therefore, all the fundamental frequencies are given by the equation

$$f_m = 261.6 \cdot 2^{m/12}$$

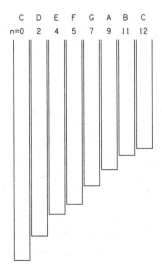

Fig. 8.7 A pan flute (pan pipe) that plays an equal-tempered major scale using pipes that are closed at one end. Except for the small end correction, the tube lengths decrease exponentially for tones of increasing frequency

where m is the number of semitones, $m = 0, 2, 4, 5, 7, 9, 11, 12$. Notice that $m = 0$ produces middle C and $m = 12$ produces a C that is an octave higher. This equation, together with the values of m leads to the eight fundamental frequencies of the scale. Each of the notes of the scale will have a set of odd harmonics that will occur naturally without any effort on our part.

To make the flute, we will use lengths of tube with an inside diameter of 1 cm (10 mm). The diameter determines the end correction. To find out the lengths of the tubes, we use the formula for a pipe open at only one end, including the end correction,

$$f_m = v/[4(L_m + L_{end})],$$

where L_m is the length of the tube that we have to cut.

To find that length, we can combine the two equations and solve for L_m,

$$L_m = v/(4 \times 261.6 \times 2^{m/12}) - L_{end}$$

or

$$L_m = 344000/(1046 \times 2^{m/12}) - 3,$$

where L_m is the tube length in mm, 344000 is the speed of sound in mm/s, and 3 mm is the end correction for a tube with an inside diameter of 10 mm.

The last equation leads to tube lengths of 326, 290, 258, 243, 216, 192, 171, and 161 mm, as shown in Figure 8.7.

Exercises

Exercise 1, 1-ft pipe

(a) You have a pipe, 1 ft long and open at both ends. What frequency do you hear if you blow across this pipe? (Note: Use the fact that the speed of sound is $v = 1130$ ft/s. You may ignore the end correction for this part and the other parts of this exercise.)

(b) You close off one end of the pipe and blow across the open end again. Show that the fundamental frequency is an octave lower than in part (a) where both ends were open. (Recall the definition of an "octave:" Frequency f_a is an octave higher than frequency f_b if f_a is twice f_b, i.e., $f_a = 2f_b$. Then f_b is an octave lower than f_a.

(c) Explain why the open–closed pipe makes a tone with only odd harmonics (1, 3, 5, ...) whereas an open–open pipe makes a tone with all the harmonics (1, 2, 3, ...)

Exercise 2, Nodes and modes

Find the mathematical relationship between the number of nodes for the standing waves in an open–closed pipe and the mode number. Compare with the relationship for an open–open pipe. (Don't count the points of zero pressure that occur at open ends.)

Exercise 3, Open–closed pipe

Draw the standing wave for the fourth mode of an open–closed pipe. How many nodes does it have? (Don't count the point of zero pressure at the open end.) What is a mathematical expression for the frequency?

Exercise 4, Open–open pipe

Draw the standing wave for the fourth mode of an open–open pipe. How many nodes does it have? What is a mathematical expression for the frequency?

Exercise 5, The end correction

In what way does the end correction depend on the ratio of the pipe diameter to the pipe length? Does the end correction tend to zero if the pipe diameter becomes smaller and smaller? Does the end correction tend to zero if the pipe length becomes longer and longer?

Exercise 6, Pipes of Pan

Pan pipes are cylindrical tubes in an array. Blowing across a tube produces a tone with frequencies corresponding to the expected modes. The tubes are stacked in pairs, with one tube open–open and the other open–closed. Blowing across the pair of tubes creates two tones, with fundamental frequencies that are an octave apart. The two tubes of a pair are almost the same length, but not quite. Which of the two should be slightly longer? [Hint: Remember that the end correction is applied to open ends only.]

Exercise 7, The organ pipe is too long!

According to Appendix C, a massive pipe organ has a pipe that sounds a frequency of 16 Hz. That is the fundamental frequency of the lowest note. (a) Show that the formula $f_1 = v/(2L)$ implies that this pipe needs to be 10.75 m long. (b) Show that this length is more than 35 ft. (c) Unfortunately, you are confronted with an installation where the vertical space available for the pipe organ is only 30 ft. What can you do to solve this problem? (d) What do you sacrifice with this solution?

Exercise 8, Boston Symphony Hall

The front of the famous Boston Symphony Hall features 48 exposed organ pipes, and 33 of them appear to have the same length! How can the designer make organ pipes that look that way and still have the pipes play different notes?

Exercise 9, How to make the pipe more complicated

The mode frequencies for a stretched string depend on the string length, the mass density, and the tension. By contrast, the mode frequencies for air in a pipe depend only on the pipe length—much less complicated. What could be done to make the acoustics of a pipe seem more analogous to the acoustics of a string?

Exercise 10, Popping a pipe

Imagine that you have a piece of thin plumbing pipe, about 25 cm (10 in.) long and open at both ends. You can make a tonal sound by striking an open end on your palm (first remove burrs from the end of the pipe). Alternatively you could put a finger into the pipe and pop it out like a cork. Both methods shock excite the modes of the pipe. Do you expect that both methods lead to the same pitch?

Exercise 11, The clarinet

The clarinet is a woodwind instrument with a cylindrical bore about 60 cm long. When all the tone holes are closed, it resembles an open–closed pipe. Calculate the lowest frequency for the clarinet. Compare with the lowest frequency given in Appendix C.

Exercise 12, Percussive bass

You want to make a bass musical instrument from PVC pipe. The pipe should produce the musical note C2 with a frequency of 65.4 Hz when you slap an end with a wet sponge. You have PVC pipe with an inside diameter of 1.5 in. How long should you cut the pipe? Remember the end correction for the (only one) open end.

Exercise 13, Slide whistle

The slide whistle is made from a cylindrical tube that is closed at one end by a stopper. The stopper can be retracted to make the tube longer. When the stopper is fully retracted, the inside length is 287 mm. The rod that retracts the stopper has a total travel distance of 232 mm. (a) Show that the lower and upper frequencies are about 300 and 1,560 Hz. (b) The manufacturer claims that the range of the whistle is more than two octaves. Do you agree?

♠

Chapter 9
Fourier Analysis and Synthesis

The concepts of Fourier analysis and synthesis come from an amazingly powerful theorem by mathematician J.B.J. Fourier (1768–1830). Fourier's theorem states that any waveform is just a sum of sine waves. This statement has two implications, Fourier synthesis and Fourier analysis.

Fourier *synthesis* is something that you already know about from your study of vibrations and waves. You already know that a complicated vibration can be created by adding up simple harmonic motions of various frequencies. You already know that a complex wave can be synthesized by adding up sine waves. Fourier's theorem says that by adding up sine waves you can create any waveform your heart desires, *any waveform at all*.

The opposite of synthesis is *analysis*. In Fourier analysis one begins with a complex wave and discovers what the sine waves are that make it. Just as a chemist analyzes a material to discover how much of what elements are present, the acoustician analyzes a complex tone to discover how much of what sine frequencies are present. Fourier's theorem also says that any complex wave can be analyzed in *only one way*. There is only one set of sine wave frequencies, amplitudes, and phases that can come out of the analysis. For that reason, we can make a better statement of the theorem, "Any waveform is a *unique* sum of sine waves."

To summarize: In Fourier synthesis one begins with an amplitude spectrum and phase spectrum and uses those two spectra to generate a complex wave. The spectra are functions of frequency. The complex wave is a function of time.

In Fourier analysis one begins with a complex wave that is a function of time. The results of the analysis are amplitude and phase spectra, and these spectra are functions of frequency.

The following examples give (1) an equation for a wave (or signal), (2) a graph of the wave as a function of time, and (3) graphs showing the amplitude and phase spectra.

W.M. Hartmann, *Principles of Musical Acoustics*, Undergraduate Lecture Notes in Physics, DOI 10.1007/978-1-4614-6786-1_9,
© Springer Science+Business Media New York 2013

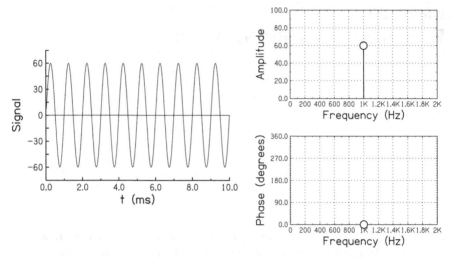

Fig. 9.1 A sine wave has the simplest possible spectrum

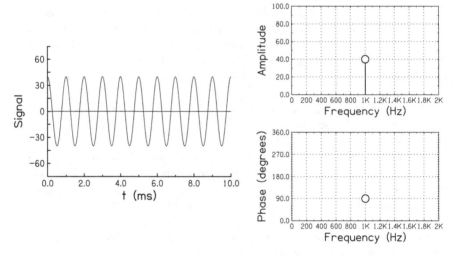

Fig. 9.2 Another sine wave. This one starts at a peak because of the 90-degree phase

9.1   The Sine Wave

The simplest wave has a single component. That makes it a pure sine wave. A sine signal with an amplitude of 60 units, a frequency of 1,000 Hz, and and a phase of zero shown in Fig. 9.1 and is given by the equation

$$\text{Signal} = 60\ \sin(360 \cdot 1000\,t + 0). \tag{9.1}$$

A variation on the 1,000-Hz signal is a sine signal with an amplitude of 40 units, and a phase of 90°, as shown in Fig. 9.2. It is given by

$$\text{Signal} = 40\ \sin(360 \cdot 1000\,t + 90). \tag{9.2}$$

9.2 Complex Waves

We next consider a complex wave with two components. There is a fundamental component with a frequency of 200 Hz and an amplitude of 45 units. There is a third harmonic with a frequency of 600 Hz and an amplitude of 15 units. (You may wish to revisit the definition of harmonics in Chap. 7.) Both components have zero phase. Please notice the positions of all those numbers, 200, 600, 45, 15, 0, and 0, in Eq. (9.3) below. You can consider this equation to be a model mathematical description for all complex waves.

$$\text{Signal} = 45\ \sin(360 \cdot 200\,t + 0) + 15\ \sin(360 \cdot 600\,t + 0). \tag{9.3}$$

Phase plays a role in shaping the final waveform. Consider what happens if the phase of the third harmonic is changed to 90° as in Fig. 9.4:

$$\text{Signal} = 45\ \sin(360 \cdot 200\,t + 0) + 15\ \sin(360 \cdot 600\,t + 90). \tag{9.4}$$

Comparing the two final waveforms in Figs. 9.3 and 9.4 makes it clear that by changing the relative phase of the first and third harmonics we have changed the shape of the waveform. Curiously, this phase change does *not* change the sound of this wave. A law called "Ohm's law of phases" says that human listeners are insensitive to phase changes. For a signal with two low-numbered harmonics, like the first and third harmonics of 200 Hz, the law holds good. More about Ohm's law appears below in a discussion of the sound of periodic waves.

9.3 Periodicity

The periodicity of the complex wave with components at 200 and 600 Hz is 1/200 s. That's because after 1/200 s (5 ms) the 200-Hz component has gone through exactly one cycle, and the 600-Hz component has gone through exactly three. Both components are then ready to start over again. That's the basis of periodicity. The component at 200 Hz is the lowest frequency component. It also happens to be the fundamental. The component at 600 Hz is the third harmonic.

But sometimes the assignment of periodicity, fundamental, and harmonic numbers is not so simple. Think about a complex wave that has components at 400, 600, and 800 Hz, as shown in Fig. 9.5. After 1/400 s (2.5 ms) the 400-Hz component

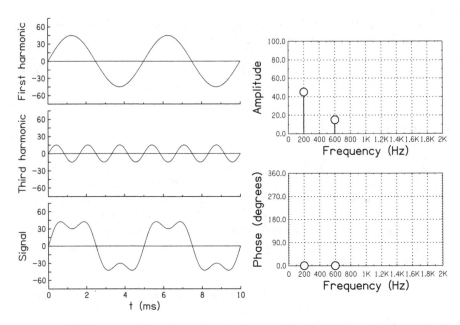

Fig. 9.3 A complex wave with two components. The two components can be seen in the amplitude and phase plots on the *right*. The figure on the *left* shows the first harmonic or fundamental (200 Hz), the third harmonic (600 Hz), and the final signal waveform. The final signal is made by summing the fundamental component and the third-harmonic component

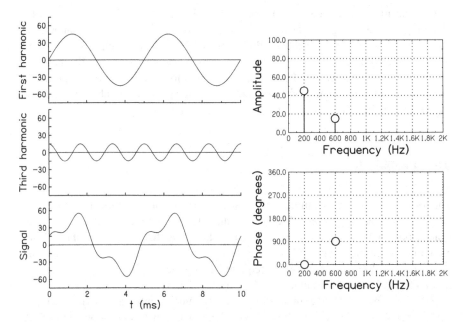

Fig. 9.4 A complex wave that is the same as in Fig. 9.3, except for the phase of the third harmonic. This phase change causes a change in the waveform but causes no change in the sound of the wave

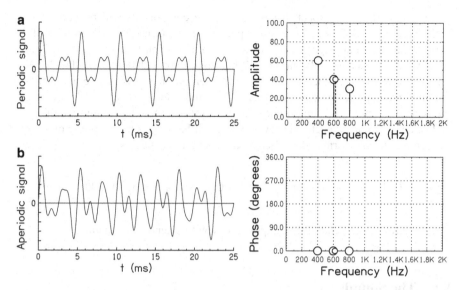

Fig. 9.5 Part (**a**) shows a complex tone with three sine components, 400, 600, and 800 Hz. The component amplitudes decrease with increasing frequency, as shown in the spectrum on the *right*, and all components have zero phase. There is a missing fundamental, with frequency of 200 Hz. The period is 1/(200 Hz) or 5 ms. Part (**b**) shows an inharmonic tone with three sine components having frequencies of 400, 620, and 800 Hz. The amplitudes and phases are the same as in part (**a**), but the frequency of the third harmonic is increased from 600 to 620 Hz, as shown by the little *circles* and *dashed lines* in the spectra. Although part (**b**) starts out looking like part (**a**), no periodicity can be seen in part (**b**)

has gone through one cycle and the 800-Hz component has gone through two but the 600-Hz component has gone through 1.5 cycles. This is not an integer number of cycles and this component is not ready to start over again. The period of this complex wave is, in fact, 1/200 s, as can be seen in Fig. 9.5a. The fundamental frequency is 200 Hz. It just happens that there is no power at 200 Hz. The fundamental is missing. What is present are harmonics 2, 3, and 4.

To find the fundamental frequency f_0 (and the period $T = 1/f_0$) when you are given a set of components like 400, 600, and 800 Hz, you find the *largest common divisor*. The largest number that divides into these three frequencies an integer number of times is 200. That makes 200 Hz the fundamental. Of course, the number 100 will divide into those three frequencies too, but 100 is not the largest number.

9.4 The Sawtooth

You will remember the sawtooth wave from the discussion of the oscilloscope. It is a function of time, and on the basis of Fourier's theorem we expect that we can create it by adding up sine wave components. Furthermore, the sawtooth is a periodic waveform, and so we expect that the sine wave components will be harmonics. If the period is 0.0025 s (2.5 ms) we expect the components to be harmonics of a 400-Hz fundamental. Figure 9.6 shows an approximation to a sawtooth. It is an approximation because only the first 12 harmonics are included, 400, 800, 1,200, ..., 4,800 Hz. All the other harmonics have been omitted. It is simple to describe the amplitude spectrum of the sawtooth. The fundamental has an amplitude of 1, the second harmonic has amplitude 1/2, the third harmonic has amplitude of 1/3, and so on. The phase spectrum is even easier. All harmonics have a phase of 180°. Therefore, the waveform is described by an equation that begins,

$$\text{Signal} = 1.\ \sin(360{\cdot}400\,t+180)+\frac{1}{2}\ \sin(360{\cdot}800\,t+180)+\frac{1}{3}\ \sin(360{\cdot}1200\,t+180)+\ldots$$

$$(9.5)$$

An exercise at the end of the chapter asks you to extend this equation.

9.5 The Sounds

This chapter has described three kinds of periodic signals. How do they sound?

The 1,000-Hz sine: Figures 9.1 and 9.2 showed 1,000-Hz sine tones with different starting phases. The starting phase affects the sound for only an instant and cannot possibly matter in the long run. A frequency of 1,000 Hz is somewhat piercing and

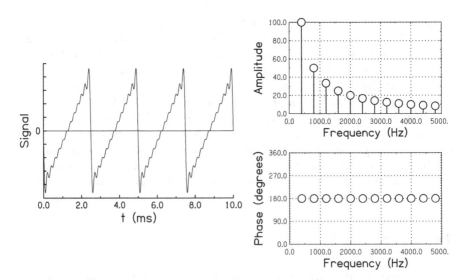

Fig. 9.6 A sawtooth with a 2.5-ms period can be approximated by adding up the first 12 harmonics of a 400-Hz fundamental

unpleasant (3,000 Hz would be worse). A 100-Hz sine tone has a "dull" tone color. The fact that this simple sine waveshape, with only a single component, can have a tone color that spans the range from dull to piercing tells you that the color of a tone depends more on the frequencies that are present in that tone than on the shape of the waveform.

The two-component complex, 200 plus 600 Hz: Figures 9.3 and 9.4 showed complex tones consisting of a fundamental component (harmonic number 1) and a third harmonic having one-third the amplitude. If you listen carefully, you can hear both harmonics separately. That is not the usual situation. Usually one does not hear the individual harmonics of a complex periodic tone, but then a typical complex tone has many harmonics, not just two. If one starts with the complex tone having harmonics 1 and 3 and then adds harmonics 2 and 4, it is no longer possible to hear the first and third harmonics separately. Instead one hears a single tone with a complex tone color.

If one starts with harmonics 1 and 3, as before, and adds harmonics 5, 7, 9, 11, etc., then once again the third harmonic tends to disappear in the crowd. A listener normally hears only a single tone. Odd harmonics like this are the basis of the square wave and triangle wave, waveforms that come out of typical function generators (recall from the end of Chap. 4).

For a typical complex tone, the fundamental frequency (reciprocal of the period) determines the pitch, and the amplitudes of the components determine the tone color or steady-state timbre of the tone. High-frequency harmonics with large amplitudes lead to a bright tone color.

Ohm's Law For periodic tones having fundamental frequencies of 100 Hz or greater and having only a dozen low-numbered harmonics, Ohm's law of phase insensitivity holds well. Ohm's law fails when there are systematic changes among high harmonics that are close together. For instance, the tone color of a waveform made from harmonics 20, 21, 22, 23 (and no others) can depend on whether the harmonics all have the same phase or whether they have random phases. These kinds of signals do not occur in nature. They can be constructed electronically for experiments in psychoacoustics. Despite failures with such specially constructed signals, Ohm's "law" holds rather well for naturally occurring signals such as the tones of musical instruments. The human ear is insensitive to the relative phases among the harmonics. By contrast, changes in the amplitude spectrum make a big difference in the sound of a tone. As a result, the phase spectrum of complex tones is not treated with the same respect as the amplitude spectrum. Most of the time we ignore the phase spectrum.

Phase-scrambled sawtooth The 400-Hz sawtooth waveform in Fig. 9.6 has a pitch of 400 Hz and a bright tone color. It could serve as the starting point for the synthesis of a trumpet tone or violin tone. We can use the sawtooth waveform to illustrate Ohm's law. Figure 9.7 shows a waveform having the same amplitude spectrum as the sawtooth in Fig. 9.6. However, the phases of the harmonics have been chosen randomly from the full range, 0–360°. Obviously the waveform looks

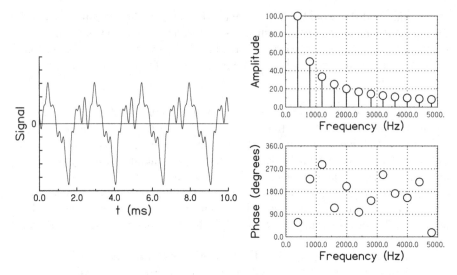

Fig. 9.7 A 400-Hz signal with 12 harmonics having the same amplitudes as the sawtooth in Fig. 9.6. However, the phases are all different from sawtooth phases

entirely different, though the period is still the same. All that is no surprise. What is surprising is that this waveform sounds just like the sawtooth waveform. Ohm's law holds well in this case.

9.6 Harmonic and Inharmonic Spectra, Periodic and Aperiodic Tones

Harmonic: Periodic A tone made out of sine waves having frequencies of 200, 400, 600, and 800 Hz has harmonic components. These sine components (also called partials) are all harmonics of 200 Hz. Because the partials are harmonic, the tone is periodic. Its period is 1/200 s and the pitch is 200 Hz, or very nearly 200 Hz.

A tone made out of sine waves having frequencies of 400, 600, and 800 Hz (see Fig. 9.5a) also has harmonic components. These components are all harmonics of 200 Hz, but the fundamental (200 Hz) is not present in the spectrum. The tone has a period of 1/200 s and the pitch is very close to 200 Hz.

Inharmonic: Aperiodic If the partials of a tone are not simple integer multiples of a fundamental, then the partials are inharmonic and the tone is aperiodic. For instance, the tone with components 400, 620, and 800 Hz has inharmonic components and the tone is not periodic. Figure 9.5b shows the waveform evolving in a complicated way. No two "cycles" are the same. The waveform does not seem to repeat itself.

But hang on a minute! Maybe this tone is periodic after all. We could say that there is a fundamental frequency of 20 Hz and the components are the 20th, 31st, and 40th harmonics of 20 Hz. [400/20 = 20; 630/20 = 31; 800/20 = 40.] The period is 1/20 s, and we might expect a pitch of 20 Hz.

In strict mathematical terms there is nothing wrong with the argument in the paragraph above, and yet the prediction about the pitch is quite wrong. That problem points up the fact that what we choose to call harmonic and periodic is determined by our auditory perceptions. Harmonic numbers 2, 3, and 4 are numbers that are small enough and close enough together that the tone with components at 400, 600, and 800 can be considered periodic. Harmonics 20, 31, and 40 are so large and so far apart that we do not perceive the complex consisting of 400, 620, and 800 as a single tone, and we do not treat it as a single periodic entity. More about this important feature of the human auditory system will appear in Chap. 13 on pitch perception.

The Tuning Fork Again Recall the discussion of the tuning fork in Chap. 3. There we added up two sine components with frequencies of 256 and 1,997 Hz. The process was just the same as adding up the components of 200 Hz and 600 Hz in Figs. 9.3 and 9.4. But there is one big difference. The mode frequencies of the tuning fork were not harmonics. Therefore, the final wave for the tuning fork in Fig. 3.4c was not periodic. Because these components were not harmonics, there was no reason to be concerned about the phases of the harmonics, and the spectrum in Fig. 3.6 showed only the amplitudes. When two components are not harmonics, their relative phase varies continuously. By contrast, the harmonics of a periodic tone are locked together. Their relative phases determine the shape of the waveform, and there may be some reason to be interested in the phase spectrum.

9.7 Filters

Possibly you have been asking yourself, "What's the point of this Fourier analysis?" "What does it matter if a waveform like a sawtooth or whatever can be analyzed into components of different frequencies, or synthesized by adding up sine tones with those frequencies?" Ultimately, the answer to these questions is *filtering*. A filter passes some things and rejects other things. For instance, a filter in a coffee maker passes coffee and rejects coffee grounds. A filter in acoustics or electronics passes certain frequencies and tends to reject other frequencies. Filtering of some sort can be found everywhere in acoustics, including the human auditory system. Because filtering is selective of *frequency*, the frequency representation made possible by Fourier analysis is crucial. The fact that filters operate in the domain of frequency is the reason that the concepts of Fourier analysis and synthesis are so important.

From your study of vibrations you have learned about resonance. Systems with natural frequencies of vibration serve as filters in that they respond strongly to only certain frequencies. If a complicated waveform is impressed upon such a system, the system will select out mainly those frequencies that are its natural frequencies. Filters are particularly evident in audio or electronics. There are lowpass filters that pass only low frequencies and highpass filters that pass only high frequencies. There are bandpass filters that act like resonant devices (resonators) because they pass only frequencies in a band near the resonant frequency.

All the World's a Lowpass Filter The consequences of Fourier's theorem are not limited to acoustics. The relationship between waveforms and frequency has enormous commercial and political consequences. The fact is, everything in the world is a lowpass filter in the sense that there is some upper limit to the frequency to which it will respond. In communication systems, the lowpass character limits the bandwidth, and this places a limitation on the rate at which information can be transmitted.

Figure 9.8a shows a data stream consisting of seven bits—ones or zeros. This data stream actually represents the capital letter "Y." It is shown here, transmitted at a rate of 10,000,000 bits per second, so that each bit lasts for 0.1 µs, which is 100 ns.

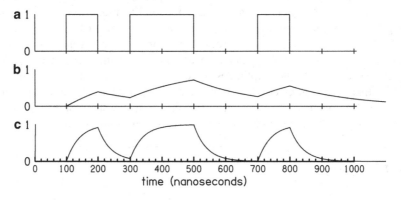

Fig. 9.8 (a) A computer code (ASCII) for the letter "Y." (b) The computer signal from (a) after it is passed through a channel with a bandwidth of only 0.8 MHz. (c) The computer signal from (a) after it is passed through a channel with a bandwidth of 4.0 MHz

Suppose we try to transmit this signal through a communications channel with a bandwidth of only 800 KHz (0.8 MHz). This channel will not pass the high frequencies needed to make a clean signal, and the received signal will look like Fig. 9.8b. It is very likely to be unreadable at the receiving end, and data are lost. In order to transmit information reliably in such a channel we need to slow down the transmission rate so that the frequencies in the spectrum are not so high. Alternatively, if we can afford it, we can buy or rent a channel with a wider bandwidth. For instance if the bandwidth of the channel is 4 MHz, the received signal looks like Fig. 9.8c. That is very likely to be a useable signal. The receiving system can recover the data without error.

Thus, by using a wider bandwidth we have been able to transmit information at the high rate.

Every communication system in the world has some bandwidth limitation. The search for wider bandwidth has led telecommunications companies to new technologies, such as optical data transmission. The scarcity of bandwidth in broadcasting has led to governmental regulation worldwide. Essentially, everywhere you look—Fourier rules!

9.8 Continuous Spectra

The spectra that appear in this chapter have all been line spectra in that they consist of discrete frequencies, like 200, 400, 600 Hz. This section introduces continuous spectra where there are no gaps between frequencies. The concept is somewhat abstract, but it can be approached from what you already know.

(1) From Fig. 9.5 you know that a wave with components at 400, 600, and 800 Hz has a short period, only 1/200 s. Its harmonics are far apart—they are separated by 200 Hz.
(2) From Fig. 9.5 you also know that a wave with components at 400, 620, and 800 Hz is less periodic. It has a longer period, 1/20 s. Its harmonics are much closer together; they are separated by 20 Hz.
(3) The tuning fork with frequencies of 256 and 1,997 Hz is even less periodic. Its period is a full second, and that caused us to say that it was not really periodic. Its harmonics (to stretch the definition of the term) would be separated by only 1 Hz—in fact, they would be the 256th and the 1,997th harmonics of 1 Hz— kind of a silly idea, but it is mathematically correct.
(4) Finally, there is a signal like the letter "Y" in Fig. 9.8. That signal is not at all periodic. It never repeats itself. It's a one-shot event. Therefore, we have to say that the period is infinite, and, to follow the mathematical logic, the separation between the spectral components becomes zero. That's the concept of the spectral continuum. Just for illustration, the amplitude spectrum is shown in Fig. 9.9. It looks continuous, not like individual discrete lines. The phase spectrum is not shown.

Exercises

Exercise 1, A complex wave

Write the equation for a wave having the first four harmonics of 250 Hz, all with 90° of phase and with amplitudes given by $1/n$ where n is the harmonic number. [Hint: Follow the form in Eqs. (9.3) and (9.4).]

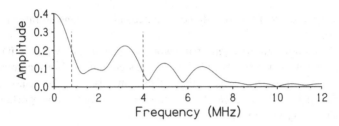

Fig. 9.9 The amplitude spectrum of the signal for the letter "Y" from Fig. 9.8. The spectrum becomes zero at 10 MHz because each pulse has a duration that is a multiple of 0.1 μs. *Dashed vertical lines* appear at 0.8 and 4.0 MHz. You can see how much spectral strength is lost if the signal is filtered to reduce the amplitude of components above 0.8 MHz. That is why Fig. 9.8b is such a poor representation of Fig. 9.8a

Exercise 2, Morphing a signal

   Figure 9.3 shows the signal described by the equation

$$\text{Signal} = 45 \, \sin(360 \cdot 200 \, t + 0) + 15 \, \sin(360 \cdot 600 \, t + 0). \qquad (9.6)$$

Use that plot to imagine the signal described by

$$\text{Signal} = 45 \, \sin(360 \cdot 200 \, t + 0) + 45 \, \sin(360 \cdot 600 \, t + 0), \qquad (9.7)$$

where the third harmonic is just as tall as the fundamental. Sketch that signal.

Exercise 3, What's the period?

(a) Use figures to show why a wave with components at 200 and 600 Hz has the period of 1/200 s. Do the phases of the components change the periodicity?

(b) What is the period of a wave with components at 200 and 500 Hz?

(c) What is the period of a wave with components at 200 and 601 Hz?

(d) What is the period of a wave with components of 150, 450, and 750 Hz?

(e) What is the period of a wave with components of 220, 330, 550, and 660 Hz?

Exercise 4, Physical vs psychological

   The discussion of the phase-scrambled sawtooth in the text says that the sawtooth waveform has a pitch of 400 Hz. Frequency is a physical quantity and it can properly be measured in Hz. Pitch, however, is a perceptual quantity, and it can only be measured by a human listener. So how can one measure a pitch in physical units of Hz?

Exercise 5, Bozo and the waveform

   Bozo explains that according to Ohm's law the amplitude spectrum determines the shape of the waveform. Therefore, the picture on an oscilloscope screen does not depend much on the phases of the components. Set Bozo straight on this matter.

Exercise 6, More components for the saw.
Equation (9.5) gives the first three components of the sawtooth wave. What are the next three?

Exercise 7, Drill.
Consider the wave given by the equation:

$$\text{Tone} = 10 \sin(360 \cdot 200\, t + 30) + 15 \sin(360 \cdot 400\, t + 60)$$

$$+\ 5 \sin(360 \cdot 600\, t - 90) + 3 \sin(360 \cdot 800\, t + 240). \qquad (9.8)$$

(a) How many harmonics are there?
(b) How many vertical lines would a spectrum analyzer show?

(c) Which harmonic has the largest amplitude?
(d) What is the amplitude of the component with a frequency of 600 Hz?
(e) What is the phase of the component with a frequency of 400 Hz?
(f) Can a phase angle really be negative (e.g., $-90°$)?
(g) Could the phase of 240° just as well be $-120°$?
(h) What is the period of this tone.
(i) If the frequencies are in units of Hertz (e.g., 200, 400 ... Hz), is it essential that parameter t be measured in seconds?

♠

Chapter 10
Sound Intensity

10.1 Pressure, Power, and Intensity

Sound waves have pressure, power, intensity, and energy. These facts became evident at the walls of Jericho some years ago. If you weren't there at the time to experience the event, then you may anyhow have experienced chest wall vibrations at rock concerts—or when listening to music in a booming car. Like light waves from a laser, sound waves have pressure, power, intensity, and energy.

At this point we mainly know about the pressure in sound waves. But pressure looks like a strange way to describe the flow of sound power because a pressure wave is both positive and negative. Normally it is negative just as much as it is positive, and its average value is zero. That seems quite different from the concepts of power, intensity, and energy. In contrast to pressure, sound power is strictly positive. It is closely related to the amplitude of the pressure (a positive number). We will refer to the pressure amplitude by the symbol A.

Focus on Intensity: I A sound wave travels outward from the source of sound, and the power in the wave gets spread out over space. If a certain amount of power is spread out over a large area, then that is a weak sound. If the same amount of power is concentrated in a small area, then the sound is strong. This concept of weak and strong is embodied in the measure called "intensity." The physical dimensions of intensity are power per unit area. Because power is expressed in metric units of watts, and area is in square meters, intensity is expressed in "watts per square meter." We refer to power and intensity by the symbols P and I respectively.

There is an important relationship between pressure amplitude and intensity: The intensity is proportional to the square of the pressure amplitude. We write

$$I \propto A^2. \tag{10.1}$$

W.M. Hartmann, *Principles of Musical Acoustics*, Undergraduate Lecture Notes in Physics, DOI 10.1007/978-1-4614-6786-1_10,
© Springer Science+Business Media New York 2013

Note that this means that if the pressure is doubled, then the intensity is increased four times. If the pressure becomes ten times larger, then the intensity becomes 100 times larger, i.e., it is increased by a factor of 100.

Proportionality

The concept of proportionality, indicated by the symbol $\propto$, is not difficult to understand. A statement of proportionality, such as Eq. (10.1) simply means that there is some constant number k that would allow us to write a true equation,

$$I = kA^2 \tag{10.2}$$

where k does not depend on either A or I. The concept of proportionality is used in mathematics to emphasize a particular functional dependence and to disregard others. Here, proportionality statement Eq. (10.1) emphasizes that intensity depends on the square of the pressure amplitude. What is disregarded in the proportionality statement is all the complexity that might be hidden in Eq. (10.2) because of constant k. For instance, k might depend on the temperature or the density of the gas where the wave is moving. The proportionality statement makes the assumption that all those quantities are unchanging (that's why k is called a "constant") as one changes the pressure amplitude and watches the change in intensity.

10.2 The Inverse Square Law

When you are farther away from a sound source, the sound tends to be less loud. That is such a familiar experience that you would imagine that there ought to be a good physical explanation for it. The most basic attempt to find such an explanation is the inverse square law.

The inverse square law refers to the *intensity* of the sound wave. It says that the intensity at a receiver depends inversely on the square of the distance from the source. If I is the intensity and d is the distance between the source and the receiver, then

$$I \propto 1/d^2. \tag{10.3}$$

The statement of proportionality in Eq. (10.3) is an abstract rule about intensity and distance. We apply this rule by making comparisons between several particular situations. For instance, let's think about a trumpet playing 1.5 m (1.5 m) away from your ear, and a trumpet playing 15 m away from your ear. If we say that I_1 is the intensity when the source is at $d_1 = 1.5$ m and I_2 is the intensity when the source is

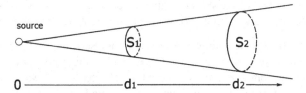

Fig. 10.1 The sound source at the left radiates power in a conical pattern. At distance d_1, the power is spread over area S_1. At distance d_2 the power is spread over area S_2. Because d_2 is twice as large as d_1, S_2 is four times larger than S_1. The sound intensity depends on how the power is spread over the area and so the intensity at distance d_2 is one-quarter of the intensity at distance d_1

at $d_2 = 15\,\text{m}$, then the square of the distance ratio is

$$(d_2/d_1)^2 = (15/1.5)^2 = 10^2 = 100. \tag{10.4}$$

The inverse square law says

$$I_2/I_1 = (d_2/d_1)^{-2} = (15/1.5)^{-2} = 10^{-2} = 1/100. \tag{10.5}$$

That means that the intensity is 100 times less when you are 15 m away from the trumpet compared to when you were 1.5 m away.

It's not hard to see why the inverse square law is true. Sound intensity is like a power density in that it is the number of watts per square meter. A wavefront from the trumpet spreads out in a beam as it leaves the horn. When the ear is only 1.5 m away from the horn, the ear intercepts the wavefront at a distance where it is spread over a small area. When the ear is 15 m from the horn, it intercepts the wavefront where it is spread over an area that is 100 times larger. That is because the surface area of the end of a beam is proportional to the square of the length of the beam.

Figure 10.1 gives another example of the operation of the inverse square law. There, the far distance is only twice as far as the near distance. Therefore, $I_2/I_1 = (1/2)^2 = 1/4$.

Spherical Source The inverse square law does not depend on having a conical radiation pattern like Fig. 10.1. The law works for any pattern. As another example, think about a pulsating sphere that radiates a sound wave equally in all directions. Such a radiator is called *isotropic*. It radiates a spherical wave. You might think of a wavefront of this wave, e.g., a pressure maximum, as the surface of a spherical balloon. The expansion of the wavefront as it moves out from the source is like the expansion of the balloon as it fills with air. As the wave expands, the power is spread out over an ever larger surface area. Similarly, as the balloon expands, the rubber is stretched over a larger area. The surface area of a sphere, S, is proportional to the square of the radius, d. The formula is $S = 4\pi d^2$. Therefore, the intensity of the sound wave measured a distance d away from the source is $I = P/(4\pi d^2)$, where

P represents the power of the source. If the power is in watts and the distance is in meters, then the intensity is in units of watts per square meter, as expected.

Application of the Inverse Square Law The inverse square law comes from a model in which there is a source, a receiver, and nothing else but air. There is nothing in this environment that would lead to reflected sound waves. Obviously, this model is an idealization because whenever a sound occurs in a room there are many reflections from the walls of the room, contrary to the model.

In fact, in many room environments the model is a very bad approximation to reality. For instance, in a classroom with a talker near the chalk board and a listener anywhere else in the room the model fails badly, and the inverse square law does not apply. It fails to describe the dependence of intensity on distance from the source.

There are some other contexts in which the inverse square law does apply. First, if one could imagine a special room where all the walls and other surfaces absorb sound completely and do not reflect any of it, then the model would apply, and the inverse square law would be valid. Such a room is called "anechoic." All surfaces in this room are covered with foam wedges about a meter long. Second, the inverse square law might apply approximately, even in a normal room if the distance between the source and receiver is small compared to the distance to any walls. Then the reflections from the walls, violating the assumptions of the model, might be small compared to the sound that travels directly from source to receiver. Third, if the source and receiver are out of doors, away from all walls, then the inverse square law applies. It applies especially well if the ground is covered with fresh snow, a good sound absorber.

Finally, the inverse square law applies in theoretical calculations in a room where the goal is to follow each sound reflection from one surface to another. This context is different from the others because it deals with sound on a microscopic time scale. Instead of thinking about the propagation of sound from a source to a receiver, this context considers the propagation from the source to the first reflection, and second reflection, and third, and so on, and finally to the receiver. The inverse square law applies to the propagation as it occurs on each step of this complicated path.

Recapitulation

Pressure: Positive and negative, measured in pascals (or micropascals).

Power: Positive only, measured in watts.

Intensity: Positive only, Power/Area, measured in watts/(meter^2).

10.3 Decibels

It's a strange fact that the pressure, power, intensity, and energy in a sound wave are not normally specified in physical units such as watts per square meter. What we do instead is to have standard reference values for all of those physical quantities and to measure a real sound wave in terms of those reference values. That's the basis of the decibel scale.

To explain how the decibel scale works, this chapter tells a long story about you and your neighbor's music. The story gives a specific example, but you should be able to generalize the concepts to other circumstances.

The story begins with a definition of a reference intensity. The standard reference for intensity is 10^{-12} W/m^2. This intensity is called the "threshold of hearing" because it is the weakest sound that humans can hear. Making measurements at your place of the music coming from your neighbor, we might find that the intensity is one million times greater than the reference intensity. That is how we refer the intensity of a sound, like music coming from your neighbor, to the threshold of hearing.

To put this in the form of an equation, if the threshold of hearing is an intensity called I_o and the music measured at your place has an intensity called I, then

$$I/I_o = 1,000,000 \text{ or } 10^6. \tag{10.6}$$

Because the power of ten that is needed to make one million is 6, we could also say that the music is six orders of magnitude more intense than the threshold of hearing. The number 6 is an exponent here. It is a power of 10. It is the power of 10 that is needed to make a million. Therefore, the number 6 is the logarithm of one million. In equation form we say,

$$6 = \log(1,000,000). \tag{10.7}$$

There is only one step remaining to describe the music in terms of decibels. We multiply by 10. In the end we say that the music *has a level of 60 decibels* or 60 dB.

This statement, that the sound has a level of 60 dB, is a little curious. It looks as though we are describing the level in absolute terms. We know, however, that this level has really been referred to the threshold of hearing. So we might have said instead, "The sound has a level of 60 dB with respect to the threshold of hearing." The point is that the threshold of hearing has been agreed upon as an international standard. Therefore, it is OK to say simply that the level is 60 dB. We must bear in mind though that levels measured in dB always are *relative* measures, even though levels with respect to the threshold of hearing are expressed as though they were *absolute*. That distinction becomes clarified in the next section.

To summarize what has gone above, we can calculate the level, L, of a sound that has intensity I by the equation

$$L = 10 \log(I/I_o), \tag{10.8}$$

where $I_o = 10^{-12}$ W/m^2.

10.4   Absolute vs Relative dB

You go to your neighbor's place and walk in. There, the music is 100 times more intense than it was when you heard it back in your own place. Therefore, its intensity is 10^8 times the threshold of hearing. The intensity is actually 10^{-4} W/m². To calculate the level in dB, we use the formula above,

$$L = 10\log(10^{-4}/10^{-12}) = 10\log(10^8) \tag{10.9}$$

$$L = 10 \times 8 = 80 \text{ dB.} \tag{10.10}$$

At your neighbor's place the level is 80 dB. That is 20 dB more than back at your place. We arrive at this difference of 20 dB by remembering that at your place the level was 60 dB, and $80 - 60 = 20$. We calculated two values on the "absolute" dB scale and subtracted one from the other.

Another Way There is another way to do this calculation that emphasizes the dB measure as a relative measure. We return to the fact that the sound intensity was 100 times greater at your neighbor's place.

We define the level at your own place as L_1 and the level at your neighbor's place as L_2. The difference is

$$L_2 - L_1 = 10 \log(I_2/I_1). \tag{10.11}$$

Because $I_2/I_1 = 100$, we have

$$L_2 - L_1 = 10 \log(100) = 10 \times 2 = 20 \text{ dB} \tag{10.12}$$

So we get the same answer, 20 dB, whichever way we do it. We say, "Relative to the sound at your place, the level at your neighbor's place is up by 20 dB."

Another way to say it is, "Relative to the sound at your neighbor's place, the level at your place is down by 20 dB." Still another way to say it is, "The barriers between you and your neighbor attenuate the sound by 20 dB." To *attenuate* means to reduce the level.

It is instructive to think about the attenuation picture algebraically. The reference sound is the sound at the source (your neighbor). Its level is L_2. The sound of interest is the sound at your place, and its level is L_1. Therefore,

$$L_1 - L_2 = 10 \log(I_1/I_2) \tag{10.13}$$

We know that the ratio I_1/I_2 is 1/100 or 10^{-2}. We calculate

$$L_1 - L_2 = 10 \log(10^{-2}) = 10 \times (-2) = -20 \text{ dB} \tag{10.14}$$

We say that relative to the source, the level at your place is -20 dB.

Examples

By studying the following examples, you ought to be able to figure out how to do decibel problems. The first example uses powers of ten that are integers and don't require a calculator to calculate logs. Examples 2–5 use powers of ten that are not integers. You should use the *log* button on your calculator to check the numbers.

Example 1(a), An easy dB problem
    If sound A is 1,000 times more intense than sound B, then $I_A = I_B \times 10^3$. The exponent is 3 and the level of A is said to be 30 dB greater than the level of B ($3 \times 10 = 30$). *viz*

$$L_A - L_B = 10 \log(I_A/I_B) = 10 \log(10^3) = 30 \text{ dB.}$$

Example 1(b), A negative dB problem
    If sound C has one tenth the intensity of sound D, then $I_C = I_D \times (1/10)$ or $I_C/I_D = 10^{-1}$. The exponent is -1 and the level of C is said to be 10 dB *less* than the level of D.

viz

$$L_C - L_D = 10 \log(I_C/I_D) = 10 \log(\frac{1}{10}) = 10 \log(10^{-1}) = -10 \text{ dB.}$$

Example 2(a), A dB problem
    Noise Y is five times more intense than noise X. What is the level difference?

Solution
    With the levels of X and Y defined as L_Y and L_X, the level difference is given by

$$L_Y - L_X = 10 \log(I_Y/I_X)$$
$$L_Y - L_X = 10 \log(5) = 10 \times 0.7 = 7 \text{ dB.}$$

We conclude that the level of Y is 7 dB greater than the level of X.

Example 2(b), The same dB problem
    Another way to compare noises X and Y is to say that noise X is one fifth as intense as noise Y. What is the level difference?

Solution
    The level difference is given by

$$L_X - L_Y = 10 \log(I_X/I_Y)$$
$$L_X - L_Y = 10 \log(1/5) = 10 \times (-0.7) = -7 \text{ dB.}$$

We again conclude that the level of Y is 7 dB greater than the level of X.

Example 3, A reverse dB problem

The level of the Chevrolet horn (C) is 5 dB greater than the level of the Ford horn (F). Compare the intensities.

Solution

$$L_C - L_F = 5 = 10 \log(I_C/I_F)$$

$$0.5 = \log(I_C/I_F)$$

$$I_C/I_F = 10^{0.5} = 3.16.$$

We conclude that the Chevrolet horn is about three times more intense than the Ford horn.

Example 4, Absolute dB

The intensity of the trombone at your ear is 4×10^{-3} W/m². What is the level?

Solution

$$L = 10 \log(4 \times 10^{-3}/10^{-12})$$

$$L = 10 \log(4 \times 10^{9})$$

$$L = 10 \log(4) + 10 \log(10^{9})$$

$$L = 10 \times 0.6 + 10 \times 9 = 96 \text{ dB}.$$

That's loud!

Example 5, Reverse absolute dB problem

A sound level meter at the back of the hall measures 103 dB during a rock concert. What is the intensity of sound there?

Solution

Let I be the intensity of interest. Then

$$103 = 10 \log(I/10^{-12})$$

$$I/(10^{-12}) = 10^{10.3}$$

$$I = 10^{10.3-12} = 10^{-1.7} = 0.02$$

i.e., 0.02 W/m². That's dangerously loud!!

Falling off a log

1. Prove that the log of 10 is 1, in other words, $\log(10) = 1$.
2. Prove that the log of 1/10 is -1, in other words, $\log(1/10) = -1$.

3. A logarithm is an exponent. It is a power of 10. For instance, the log of 17 [$\log(17)$] is the power of 10 needed to make the number 17.

    3a. Use a calculator to show that $\log(17)$ is approximately 1.23
    3b. Use a calculator to show that $10^{1.23} \approx 17$.

4. Show that because exponents add, the log of a product is the sum of the logs. To be more precise, if A and B are any two numbers, then the log of the product of A times B is the log of A plus the log of B. In other words,

$$\log(AB) = \log(A) + \log(B).$$

5. The number 170 is 10×17. If the log of 17 is 1.23, then prove that the log of 170 is 2.23.
6. The reciprocal of 17 (1/17) is about 0.0588. If the log of 17 is 1.23, how do you know that $\log(0.0588) = -1.23$?

Exercises

Exercise 1, Outdoors
   The inverse square law for sound intensity applies pretty well outdoors where the only reflection is from the ground. (a) Show that if you go three times farther away from a source of sound, the intensity goes down by a factor of 9. (b) Show that this decrease corresponds to a level decrease of 9.5 dB.

Exercise 2, The inverse first-power law

(a) Use the inverse square law and the relationship between intensity and amplitude to show that the amplitude A of a sound wave depends inversely on the first power of the distance d to the source, i.e.,

$$A \propto 1/d.$$

(b) To solve part (a) you need to take the square root of both sides of a proportionality. Show that this is a legitimate step writing the proportionality as an equation and then taking the square root of both sides of the equation (a legitimate step in this context).

Exercise 3, A spherical radiator

A spherical source radiates 10 W of acoustical power uniformly in all directions. (a) What is the sound intensity at a distance of 1 km, assuming no obstructions. (b) What is the absolute level in dB? (c) Where would one find such conditions?

Exercise 4, Life in different dimensions

You know that the inverse square law for intensity arises because an expanding wave spreads over the surface of a sphere, and the surface area is proportional to the *square* of the distance from the source. The spherical surface itself is appropriate in this case because the world is three dimensional.

In a two-dimensional world, the wave is confined to two dimensions. An expanding wave spreads over a circle, and the circumference of the circle is proportional to the *first power* of the distance from the source. In a two-dimensional world the inverse square law for sound intensity is replaced by an inverse first-power law.

How about a one-dimensional world? A tin-can telephone or the speaking tube on a ship are examples of sound waves confined to one dimension. What then is the appropriate law for sound intensity?

Exercise 5, Orchestra vs solo

Thirty violins lead to an intensity that is 30 times greater than one violin. Show that the difference is 14.8 dB.

Exercise 6, A factor of two

Show that doubling the intensity of sound leads to a level increase of 3 dB.

Exercise 7, Unequal duet

The trombone is playing at a level that is 20 dB higher than the flute. (a) Compare the intensities of trombone and flute. (b) Compare the sound pressure amplitudes of trombone and flute.

Exercise 8, Converting the equation to amplitude

Show that

$$L_2 - L_1 = 20 \log(A_2/A_1), \tag{10.15}$$

where A refers to the pressure amplitude.

Exercise 9, Loud!

Not far from the space shuttle main engines the sound level is measured to be 130 dB. Show that the intensity is $10\,\text{W/m}^2$.

♠

Chapter 11
The Auditory System

This chapter begins a new aspect of the science of acoustics, the study of the human listener. We continue to proceed in an analytical way by studying the physiology of hearing. The physiology begins with anatomy—the science of where things are in the body, what they look like, how they are connected, and how these facts give clues about physiological function.

Overall, the auditory system consists of peripheral elements and central elements. The peripheral elements are near the skull and comprise what is commonly called "the ear." The central elements are located in the brain—the brainstem, midbrain, and cortex. The peripheral part of the auditory system is shown in Fig. 11.1—a complicated figure.

11.1 Auditory Anatomy

The anatomy of the peripheral auditory system has three main divisions: (1) outer ear, (2) middle ear, and (3) inner ear, as shown in Fig. 11.2. Figure 11.2 is a simplified version of Fig. 11.1. Please try to identify the parts of Fig. 11.1 that belong to the divisions shown in Fig. 11.2.

11.1.1 The Outer Ear

The outer ear is the part that you can see. It consists of the *pinna* and the ear canal. The pinna is the fleshy, horn-like protuberance from the side of the head—often used for hanging decorations. The ear canal (*external auditory meatus*) is a duct, about 2.5 cm long, running from the pinna to the eardrum. It sometimes accumulates ear wax. The eardrum, or *tympanic membrane*, is the end of the outer ear; it is the beginning of the middle ear.

W.M. Hartmann, *Principles of Musical Acoustics*, Undergraduate Lecture Notes in Physics, DOI 10.1007/978-1-4614-6786-1_11,
© Springer Science+Business Media New York 2013

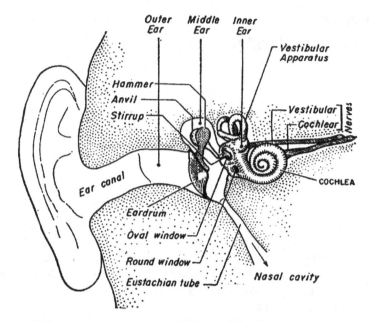

Fig. 11.1 The entire peripheral auditory system, plus the semicircular canals

Fig. 11.2 A simpler figure of the outer, middle, and inner ear

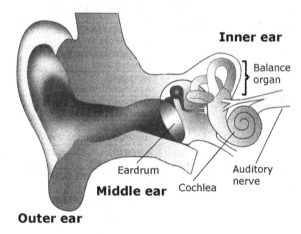

11.1.2 The Middle Ear

The middle ear consists of three bones and several muscles in a small cavity in the temporal bone. The bones act as a lever that conducts sound vibrations from the eardrum to the *oval window*. The oval window is the point of contact with the inner ear. The three bones are called *ossicles*, meaning "little bones." They are appropriately named because they are the smallest bones in the body. From the eardrum to the oval window, the bones are, in order, the *malleus*, the *incus*, and the *stapes*—translated: the hammer, the anvil, and the stirrup.

Fig. 11.3 The middle ear has three bones, hammer, anvil, stirrup, connecting the eardrum (tympanic membrane) to the oval window. The stirrup (stapes) pushes on the oval window

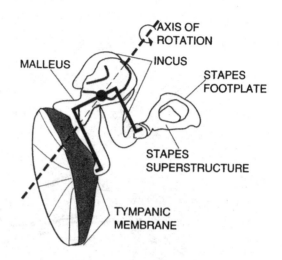

The middle ear cavity is normally filled with air, but it is sealed off at one end by the eardrum and at the other by the oval window and round window. When the outside pressure changes (perhaps because you are going up in an airplane), there would be painful pressure on the eardrum if it were not for the *eustachian tube* that connects the middle ear cavity to the nasal cavities, which, in turn, are open to the outside air. The eustachian tube allows the pressure in the middle cavity to become equal to the atmospheric pressure outside. This is a good thing. However, infections, such as a common cold, that affect the nasal cavities can make their way up to the middle ear cavity. This is a bad thing. Infants are particularly susceptible to middle ear infections. When an infection does reach the middle ear, it causes the eardrum to be inflamed. Because the eardrum is translucent, it is possible to see this inflammation by looking down the ear canal with an otoscope. That is what the physician is looking for when he or she inspects your ear canals.

The last bone in the chain, the stapes, presses on the oval window and transmits vibrations to the inner ear (Fig. 11.3).

11.1.3 The Inner Ear

The inner ear is where the real action takes place. It is possible to hear without an outer ear. It is possible to hear without a middle ear. But it is not possible to hear without an inner ear. The inner ear is responsible for converting acoustical signals into signals that the brain can understand.

The inner ear is a cavity in the skull's temporal bone called the "cochlea." It is curled up into a snail shape and divided into three canals by two membranes as shown by the cross section in Fig. 11.4. The canals are filled with fluid, not unlike sea water. The main membrane is the *basilar membrane*. The other membrane is *Reissner's membrane*, and it is so light as to be unimportant mechanically. It does

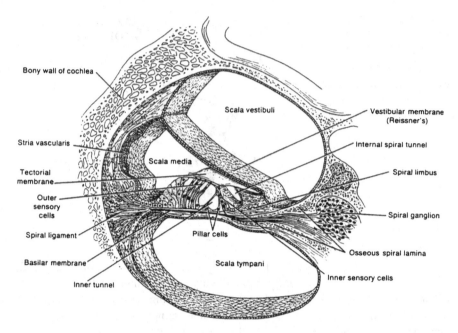

Fig. 11.4 A cross section of the cochlea showing the three canals

not affect the way the fluids move in the cochlea. But Reissner's membrane *is* important electrically. It separates the ions in the canal called "scala media" from the ions in the *scala vestibuli.* The differences in ionic concentration provide the energy source for the action of the hair cells. The ionic differences serve as the ear's battery—a source of electrical energy.

A pressure pulse from the oval window travels down the duct made from scala vestibuli and scala media to the end of the snail shell cavity where there is a small opening called the "helicotrema." There the pulse can turn around and come back on the other side of the basilar membrane through the scala tympani. At the end of the scala tympani is the *round window*, another membrane that acts as a pressure relief mechanism. The response to a pressure pulse at the oval window is shown in Fig. 11.5 by dashed lines.

Motion of the fluids in the canals causes the basilar membrane to move. On the basilar membrane is the *Organ of Corti* (Fig. 11.6), which is filled with hair cells that play a most vital role in hearing. The hair cells are transducers that convert mechanical motion into neural impulses. The neural impulses are electrical spikes that are essentially the language of the brain.

Communication within the nervous system is by means of such electrical *spikes*. Impulses from the hair cells travel along the *auditory nerve* to higher centers. The auditory nerve is sometimes called the "eighth (VIII-th) cranial nerve," or "cochlear nerve."

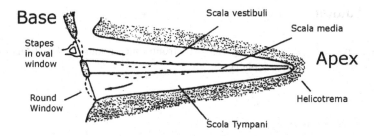

Fig. 11.5 A lengthwise X-ray cartoon of the inside of the cochlea as though it were uncoiled. What you see here is actually coiled up two and a half turns in the human head. The *dashed lines* show the response to a positive pressure pulse

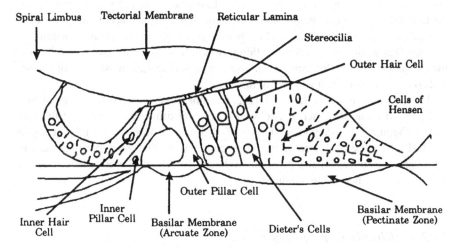

Fig. 11.6 The organ of Corti, sitting on the basilar membrane

11.1.4 The Semicircular Canals

The *semicircular canals* are contiguous with the cochlea and share the same fluids. However, they have nothing to do with hearing. They are the origin of our sense of balance (*vestibular system*). In fact, the semicircular canals work rather like the cochlea in that motion of fluids result in neural spikes that are sent to higher centers of the nervous system. However, the mechanism of the semicircular canals is sensitive only to the low frequencies of whole head motion, mostly below 10 Hz. The cochlea is sensitive to the high frequencies of the auditory world, 20–20,000 Hz.

The neural messages from the semicircular canals also travel on the VIII-th nerve. Diseases, such as Meniere's disease, that affect the vestibular system also affect the auditory system. Sometimes the cochlea and semicircular canals are jointly referred to as the "inner ear," but only the cochlea merits further study as part of the hearing system.

11.2   Auditory Function

Knowing the auditory anatomy gives you strong clues about the functions of the divisions of the auditory periphery.

11.2.1   Outer Ear Function

The pinna gathers sound and directs it into the ear canal. Its peculiar shape gives it an uneven frequency response. It captures some frequencies better than others, though this unevenness only has a big effect on high frequencies above 5 kHz. Furthermore, the frequency response depends on the direction that the sound is coming from. That is not the sort of behavior one would expect from a high-fidelity sound transmitting system. That would seem to be a bad thing. However, we humans have learned to use the asymmetry of this frequency response to help us localize the sources of sound, and so it turns out to be a good thing in the end.

The ear canal is about 2.5 cm long and it looks like a tube that is open at one end and closed at the other. It is closed by the ear drum. An exercise at the end of the chapter will ask you to show that such a tube has a resonance at about 3,400 Hz. Because of this resonance one might expect that the auditory system is most sensitive to frequencies between 3,000 and 4,000 Hz. This is actually true, as will be seen in Chap. 12 on loudness.

11.2.2   Middle Ear Function

Sound waves in the *ear canal* are pressure waves in the air inside that canal. Sound waves in the *inner ear* are pressure waves in the fluids of the cochlea—similar to seawater. The function of the middle ear is to form an efficient coupling of the waves in the air to the fluids. If it were not for the bones of the middle ear, most of the waves in the air of the ear canal would be reflected from the denser fluids of the cochlea, and little of the sound energy would be coupled into the motion of the fluids. In fact, the mismatch between the air and fluids is so bad that only 1 % of the sound energy would be transmitted from the outer ear into the inner ear. The middle ear solves that problem.

At the same time, the middle ear can reduce the efficiency of the coupling between outer and middle ear by using two muscles, the *stapedius muscle* and the *tensor tympani*. These muscles contract to make the coupling less efficient in the presence of loud sounds. This serves as a form of automatic volume control to protect the delicate inner ear. Unfortunately this is a system that is incompletely evolved and is rather slow in its action. It is too slow to react to impulsive sounds like the banging of a hammer, the explosion of gunfire, or the beat of a drum. However, it works well on sustained amplified guitar.

The muscles also contract to make the middle ear coupling less efficient when you begin to vocalize. This defends your inner ear against your own voice.

11.2.3 Inner Ear Function

As described above, the function of the inner ear, or cochlea, is to convert sound vibrations into neural impulses that the brain can understand. The action of the cochlea is one of the most fascinating stories in physiology. The cochlea is an incredibly sensitive mechanical system that gains sensitivity and sharp frequency tuning from internal electromechanical feedback. It appears that the feedback coupling can also be modulated by higher centers of the system. The cochlea is responsible for encoding signals sent to the brain, and this encoding is done in two very different ways.

Place Encoding The basilar membrane is caused to vibrate by the vibrations of the fluids in the cochlea. The basilar membrane is heavier near the apex than at the base and it vibrates maximally at different places for tones of different frequency.

A high-frequency sine tone causes vibration of the membrane near the base. (Note where the base is in Fig. 11.5.) There is little vibration anywhere else. As a result, the hair cells near the base are caused to fire, and hair cells elsewhere on the membrane are silent. This specific activity of the hair cells near the base provides a way to encode the fact that the frequency is high. The information about the *location* of active hair cells is transmitted to the brain because each hair cell is connected to higher auditory centers by about ten nerve fibers in the auditory nerve that are dedicated to that hair cell. Therefore, the brain can recognize the frequency of a tone just by knowing which hair cells are active and which are inactive.

A low-frequency sine tone causes vibration of much of the basilar membrane, but the maximum vibration occurs near the apex. The brain can recognize a low-frequency tone because hair cells near the apex are the most active. Thus, every location on the basilar membrane is best excited by a particular frequency; this is called the "characteristic frequency" of the location. A plot showing the characteristic frequency of a place on the basilar membrane for each location is given in Fig. 11.7.

The pattern of vibration is not symmetrical on the basilar membrane. The peak of the pattern is given by Fig. 11.7, but the pattern decreases abruptly on the apex side and decreases slowly on the base side when the tone has a moderate or high level. Figure 11.8 shows the responses of the basilar membrane to two separate tones with different frequencies.

For a complex tone with many frequencies, the low-frequency components cause activity in hair cells near the apex and the high-frequency components cause activity

Fig. 11.7 Frequency map of the basilar membrane according to the Greenwood equation (see Exercise 8). The *horizontal axis* shows the location of the peak of the vibration pattern for a sine tone whose frequency is given on the *vertical axis*

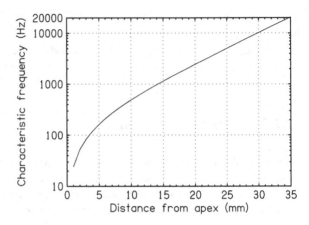

Fig. 11.8 A cartoon showing the vibration amplitude of the basilar membrane for two separate tones. A 1,000-Hz tone leads to a vibrational peak at 14.1 mm, and a 1,300-Hz tone leads to a peak at 15.8 mm. The patterns extend more toward the base (high-frequency side) than to the apex

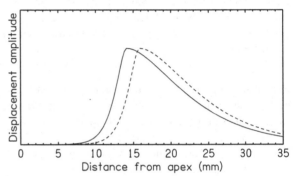

in hair cells near the base. Thus, the basilar membrane performs a spectral analysis (Fourier analysis) of the incoming wave. Different neurons transmit the signals from the various sine components. The analysis by frequency that begins in the cochlea is maintained at higher levels of the auditory system.

Timing Encoding   A hair cell encodes information about the frequency of a tone in yet another way, by synchronizing the pulses it generates with the tone itself. When a hair cell is excited by a tone it fires more frequently, but the spikes are not generated at random times. Instead, the hair cell tends to produce a spike at a particular phase of the sine waveform as shown in Fig. 11.9. As a result, the firing tends to preserve the period of the waveform. The time between successive spikes might be equal to a period, or two periods, or three, or four, ... Therefore, frequency information is passed to the brain by means of the regular firing of the neurons. In order to transmit information in this way the neurons must be able to synchronize to the signal. The nervous system cannot keep up with a signal having a frequency greater than 5,000 Hz, but it can faithfully follow a signal with a frequency of 2,000 Hz and below. More about time encoding and neural synchrony appears in Chap. 13 on pitch perception.

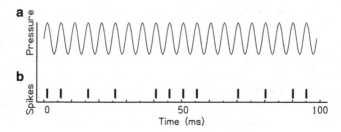

Fig. 11.9 Part **a** shows the pressure in the cochlear ducts caused by a 200-Hz sine tone. Part **b** shows the electrical spikes recorded from an auditory nerve that is tuned to a frequency near the sine tone frequency. The spikes occur on peaks of the pressure waveform, though not on every peak. In this way, the spikes are synchronized with the sine tone

How Do We Know About the Cochlea? For all its complexity, the cochlea is extremely small. It is deeply embedded in the hardest bone in the body and is also very delicate. Perhaps you are wondering how anyone could learn about the action of the cochlea, about the place encoding and timing encoding, in view of the difficulties of studying it. For instance, how could one do experiments on the cochlea without seriously affecting its action? How does one know that the experiments show the real action? Questions like these frequently arise in many different sciences. Fortunately, in the case of the cochlea there is a multi-pronged approach that has been successful.

First are experiments that open the cochlea. The opening allows the experimenter to view the basilar membrane with a high powered microscope, or to insert electrical or optical probes, or to place radioactive particles on the membrane to measure velocity [footnote 1]. These experiments can demonstrate the tuning of the cochlea, from base to apex as described above, but obviously one must be concerned about damage to the cochlea that affects the very process that one is trying to measure.

Second are experiments that allow an experimenter to infer the action of the cochlea without actually opening the cochlea to see it directly. The key to these experiments lies in the fact that individual nerve fibers in the auditory nerve are connected to particular hair cells on the basilar membrane. It is possible to insert an electrical probe into the auditory nerve and pick up the firing of a *particular* nerve, connected to a particular place on the basilar membrane. That kind of experiment demonstrates the sharp frequency tuning of the cochlea. It also shows the synchrony between the neural spikes and the signal that is presented. Experiments probing the auditory nerve also give evidence of the active feedback mechanism that is responsible for the high sensitivity and sharp tuning of the cochlea. Because the feedback process is active, it requires oxygen from the blood. Depriving an experimental animal of oxygen while monitoring the tuning and sensitivity via the auditory nerves shows that the tuning becomes much broader and the system becomes less sensitive. The cochlea seems to be acting much like the cochlea of a person whose hair cells have been damaged by a drug. Restoring normal oxygen brings back the normal tuning and sensitivity.

Third are experiments that measure cochlear emissions. Because the cochlea is electromechanically active it actually generates acoustical signals of its own. These signals propagate back along the basilar membrane, back through the middle ear, and back into the ear canal where they can be measured by microphones placed in the canal. The signals that return from the cochlea in this way are delayed compared to sounds that excite the cochlea into action. The delay makes it possible to recognize the emissions and to study them. While the open-cochlea and auditory-nerve experiments can only be done on animals, the experiments that measure otoacoustic emissions can be done on human beings. They are even part of infant screening tests.

11.2.4   Beyond the Cochlea

The hair cells on the basilar membrane in the cochlea convert sound waves into neural impulses, or electrical spikes. Higher stages of the auditory system process these impulses. The auditory nerve transmits impulses from the hair cells to a collection of neural cells called the cochlear nucleus. The cochlear nucleus is in the brainstem, part of the brain just above your neck. After the cochlear nucleus comes half a dozen stages of neural processing which include millions of neurons and an elaborate cross-linking between left- and right-hand sides. The processing proceeds by excitation, where a spike in one neuron causes a spike in another, and by inhibition where a spike on one neuron inhibits spike generation by another. The auditory pathway ascends through the processing stages to the auditory cortex, located near your left and right temples. In addition to the ascending pathway are descending pathways by which the higher auditory centers send neural messages to lower stages, even into the cochlea! The descending pathways are a form of feedback by which the processing of lower centers is controlled by the higher. The complexity of the auditory nervous system has kept psychoacousticians and auditory physiologists busy for years. They are still hard at work. One aspect of the system that seems clear is that frequency tuning, originally established by the basilar membrane is retained throughout the higher stages. At every stage, it is possible to locate a 1,000-Hz place, differently located from, say, a 200-Hz place.

11.2.5   Hearing Impairments

Ringing in the Ears   Ringing in the ears is an everyday name for *subjective tinnitus,* a sound that you hear that does not have an external physical origin. Many people have the experience that while doing nothing in particular, they suddenly hear a high-pitched ping. Often it is lateralized to one side of the other. This sound lasts a few seconds and then goes away. Such experiences are minor curiosities and so far as one knows they have no particular implications for the health of the ear or brain. Unfortunately, there are many people for whom the sound does not go away. It is a persistent tone, or buzzing, or hissing sound—usually tonal, meaning that it has a pitch. For such people, and they are perhaps as much as 20 % of the population, tinnitus can become a serious disturbance, significantly reducing the quality of life.

Tinnitus is so common that you might expect it to have a simple explanation. Surprisingly, there is no simple single explanation. Tinnitus can originate in the external ear, with an accumulation of ear wax, or in the middle ear with infection or otosclerosis, wherein a bony mass grows within the middle ear cavity and impedes the motion of the stapes. It can originate in the inner ear, especially in connection with hair cell loss caused by an acoustical trauma such as a nearby explosion. It can apparently originate in the auditory nerve, especially in the case of an acoustic neuroma, a slow-growing tumor that presses on the nerve. It may originate in the brain.

Tinnitus can be caused by common drugs, aspirin or non-steroidal anti-inflammatory drugs such as ibuprofen, or by aminoglycocides, potentially highly ototoxic antibiotics used to treat systemic bacteriological infections. Drug-induced tinnitus often ceases when the drug regimen is changed.

In some cases, tinnitus can be a warning of other disorders such as anemia, an aneurysm, or a tumor, but most of the time it seems that tinnitus has no other health implication beyond the annoyance of the tinnitus itself. There is no recognized cure for the problem. People who suffer from tinnitus are encouraged to avoid loud noises, which may aggravate the problem. Paradoxically many sufferers gain relief from masking noise, either from a loudspeaker or from a hearing aid. For most tinnitus patients, masking noise that is 14 dB above threshold is adequate to obscure the tinnitus. In some cases, masking noise may even provide relief for minutes, hours, or days after it is turned off. Programs that use biofeedback or relaxation techniques have proved helpful to some. For some individuals nothing helps, and they are required to learn to live with the problem. In summary, although tinnitus would seem to be a simple and common phenomenon, it turns out to be surprisingly complicated.

Hearing Loss and Hearing Aids Hearing loss can be classified as one of two types: conductive loss or sensorineural loss. Conductive loss occurs when the normal acoustical path to the inner ear is blocked in some way. An accumulation of ear wax can block the ear canal. Otosclerosis can prevent normal conduction of sound by the ossicles. Conductive loss can often be cured surgically. If not, conductive loss responds well to hearing aids because hearing aids essentially solve the problem of conductive loss, namely insufficient acoustical intensity arriving at the inner ear. A hearing aid consists of a microphone to pick up the sounds, an amplifier which requires battery power, and a tiny loudspeaker that fits in the ear canal. The amplifier may include filtering and other signal processing to try to tailor the hearing aid to the individual patient.

Sensorineural loss is usually a deficit in the inner ear. Normal hair cells are essential for normal hearing. Unfortunately, hair cells are vulnerable to ageing, to intense noises, to ototoxic drugs, and to congenital abnormalities. The classic case of impaired hearing results from poorly functioning hair cells. Because the hair cells are responsible for sensitivity, tuning, and timing, people who are hearing impaired may have deficits in some or all of these functions.

There is a tendency to think about hearing aids as analogous to eyeglasses, and for people with conductive hearing loss the analogy is not bad. But for the large number of people with sensorineural loss the analogy is not accurate. For the great majority of visually impaired individuals, eyeglasses (or contact lenses) solve the problem. The optical defects of an eyeball that does not have exactly the correct shape are readily compensated by passing the light through corrective lenses—end of story. Hearing impaired individuals do not normally gain the same kind of benefit from hearing aids. Hearing impairments, resulting from poor hair cell function, are as complicated as the hair cells themselves. A better analogy with the visual system would compare abnormal hair cells in the cochlea with an abnormal retina in the eyeball.

Cochlear Implants In order to benefit from hearing aids, an impaired person needs to have some residual hair cell activity. Some people however are totally deaf with no hair cell function at all. But such people often still have functioning auditory nerves. The nerve endings come to the cochlea in the normal way, but they don't find normal hair cells there to excite them.

Thousands of people who are totally deaf because of hair cell loss have regained some auditory function by the use of cochlear implants. The implant consists of a microphone that picks up sounds like a hearing aid, but the resemblance to hearing aids stops there. Instead, an electrode is inserted in the cochlea via surgery through the middle ear's round window. The electrode excites the auditory nerve directly by an electrical representation of the wave from the microphone.

Early cochlear implants were not a great success. Although they restored some sense of hearing, they did not make it possible to understand speech. These implants had only one electrode. From the place-encoding principle, however, it is known that nerve endings near the base of the cochlea normally receive high-frequency signals and nerve endings near the apex receive low-frequency signals. A modern cochlear implant has many electrodes, as many as 24, all along its length, encoding different frequency regions. Sounds with different frequencies can be sent to the correct fibers of the auditory nerve. Thus, the implant itself performs the Fourier analysis that is normally performed by the basilar membrane. Implants like this, retaining the normal frequency analysis of the cochlea, have enabled deaf persons to understand speech, even over the telephone where they cannot gain anything from lip reading.

Cochlear implants and the encoding of information sent to the electrodes have been optimized for speech. The sounds perceived by implanted patients are very different from normal speech—they have been compared to the quacking of ducks. Nevertheless, people who learned to speak and to understand speech prior to deafness have been able to make the transition and decipher the quacking. Initially, implants were restricted to people who were post-lingually deaf because it was reasoned that there was no point in providing degraded speech information to a patient who had never heard any normal speech in the first place.

Eventually, implants were tried on deaf children, and the results were surprisingly good. Further, it was found that the younger the child, the better the results! Now, infants are implanted, often so successfully that the child learns to speak and understand essentially like a normal hearing child. An important philosophical take-home message from that experience is that those of us who have normal hearing and perceive speech in the usual way, do so because our brains have learned to interpret the signals sent by the periphery. Marvelous though the peripheral auditory system may be, there is no need for it to encode signals in one specific way or another; the young brain can learn to cope with a wide variety of inputs from the periphery.

There are some drawbacks to cochlear implants. The present encoding strategy that has proved to be so successful in encoding speech removes timing information from the input signal and the patient receives no normal timing encoding. This has the effect of eliminating most musical pitch information. Implantees, even young ones, don't perceive music normally at all. Also, the surgery that implants electrodes wrecks whatever normal features there may be in the cochlea. If someday a drug is found that causes normal hair cells to grow and restore hearing to deaf people, implantees are unlikely to benefit. For the present, cochlear implants have given normal speech perception to individuals who would otherwise be as deaf as bricks. Implants represent a strikingly successful collaboration between medical practice and fundamental research in human hearing.

Cochlear Nucleus Implants A cochlear implant can restore hearing to individuals who have a functioning auditory nerve, but what if there is no functioning auditory nerve? The obvious answer is to put an implant into the next stage of auditory processing, the cochlear nucleus. A procedure that has recently been developed places electrodes in the cochlear nucleus. Although the concept of electrodes implanted in the brain stem may seem spooky, it's not the only bionic brain technology. Electrodes can be implanted in the thalamus to control the tremors of Parkinson's disease.

Exercises

Exercise 1, The ear canal as a pipe

(a) Given the description of the ear canal in the section on the outer ear, show that the resonant frequency of the ear canal (first mode) is about 3,400 Hz.
(b) What is the frequency of the second mode for air in the ear canal?

Exercise 2, Inside the canals

The speed of sound in cochlear fluid is about five times the speed of sound in air. If the cochlear duct is 35 mm long show that it takes a sound pulse about 40 μs to go from the oval window, down the duct, and back to the round window opening to the middle ear. However, the mechanical traveling wave on the basilar membrane is much slower. That wave takes about 10 ms to go from the base to the apex.

Exercise 3, Acoustic reflex

The action of middle ear muscles in response to a loud sound is called the "acoustic reflex." Describe loud sounds that you expect to trigger the action of the acoustic reflex.

Exercise 4, The "place" model of hearing.

(a) Use Fig. 11.7 to find where the basilar membrane most excited for a frequency of 440 Hz.
(b) Where are the hair cells that are most activated by the second harmonic of 440 Hz?

Exercise 5, In sync

If you were able to probe the firing of a neuron in the VIII-th nerve, what time intervals between spikes would you expect to see if the ear receives a 440-Hz tone?

Exercise 6, Place encoding

As human beings age, we lose the ability to hear high frequencies. This effect is attributed to the loss of hair cells. Where on the basilar membrane are these hair cells?

Exercise 7, Tone-on-tone masking

Intense tones of one frequency interfere with the perception of weaker tones of other frequencies. Given what you know about the mechanics of the basilar membrane, which to you expect to reveal more interference?

(a) An intense 1,000-Hz tone and a weak 1,300-Hz tone.
(b) An intense 1,300-Hz tone and a weak 1,000-Hz tone.

See Fig. 11.8 for a cartoon showing basilar membrane displacement for these two frequencies in the case of equal levels.

Exercise 8, The Greenwood equation.

The mapping of characteristic frequency (f_C) vs distance (z) along the basilar membrane was plotted using an equation published in 1961 by Donald Greenwood. This equation says,

$$f_C = 165\,(10^{az} - 1), \qquad\qquad (11.1)$$

where distance z is measured from the apex, and constant a is 0.06 inverse millimeters. Use this equation to show that the frequencies corresponding to 5, 10, 15, 20, 25, and 30 mm places are, respectively, 164, 492, 1,146, 2,450, 5,053, and 10,246 Hz.

Exercise 9, Inverse Greenwood

(a) Invert the Greenwood equation to show that the distance along the basilar membrane as a function of characteristic frequency is given by

$$z = \frac{1}{a} \log_{10}(1 + f_C/165). \tag{11.2}$$

(b) Use Eq. (11.2) to work Exercise 4(a) more precisely.

Exercise 10, The cochlear nucleus implant

Cochlear implants have multiple electrodes dedicated to different frequency bands. In this way they simulate the action of the basilar membrane—analyzing sound by frequency and sending different bands along different neural channels. Would you expect that the cochlear nucleus implant would also have multiple electrodes so that multiple frequency bands can be transmitted?

Exercise 11, The cochlear implant

The insertion of a cochlear implant is shown in the figure below. (a) Where is the conversion from acoustical to electrical? (b) What parts can be seen by an outside observer? (c) How many turns of the cochlea have been implanted?

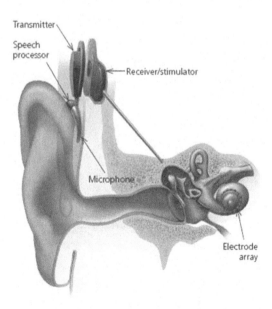

Footnote 1, Mössbauer effect: Radioactive nuclei like iron with an atomic mass of 57 units decay by emitting gamma radiation. The frequency of this radiation is amazingly precise, one part in ten billion. Because the frequency is determined so precisely, it is possible to measure small velocities of the radioactive particles by measuring the Doppler shift in the frequency of the radiation. The effect that makes this possible is known as the Mössbauer effect.

♠

Chapter 12
Loudness Perception

Loudness is a familiar property of sound. If you are listening to the radio and you want to increase the loudness you turn up the volume control, i.e., you increase the intensity. Clearly loudness is closely related to sound intensity. But loudness is not the same as intensity. Sound intensity is a physical quantity—it can be measured with a sound level meter. Loudness, by contrast, is a psychological quantity—it can only be measured by a human listener. The purpose of this chapter is to introduce the important elements of sound that contribute to the perception of loudness.

12.1 Loudness Level

On physical grounds alone you can expect that loudness should be different from intensity because the ear does not transmit all frequencies equally, i.e., it does not have a flat frequency response. The outer ear, with its complicated geometry, has several overlapping resonances including the important ear canal resonance near 3 or 4 kHz. Therefore, the magnitude of the input to the nervous system does not depend on intensity alone, the frequency also matters.

The first step on the road to a scale for loudness is a measure of intensity that compensates for the effect of frequency on the loudness of sine tones. Such a measure is the loudness level, and it is expressed in units of phons (rhymes with "Johns"). By definition, any two sine tones that have equal phons are equally loud. The curves of equal phons are called equal-loudness contours. A set of equal-loudness contours is shown in Fig. 12.1.

The contour concept is not unusual. You are probably familiar with contour maps in geography where the vertical and horizontal axes represent north–south and east–west in the usual way, and lines on the map represent equal-elevation contours. Thus one line might be the 100-m elevation line while the next line might be the 110-m line. Equal-loudness contours are like that, except that instead of equal elevation, the lines are equal loudness.

W.M. Hartmann, *Principles of Musical Acoustics*, Undergraduate Lecture Notes
in Physics, DOI 10.1007/978-1-4614-6786-1_12,
© Springer Science+Business Media New York 2013

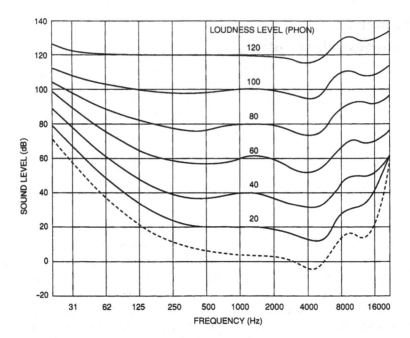

Fig. 12.1 Equal loudness contours come from loudness comparisons made by average human listeners with normal hearing for sine tones of different frequency and sound level. The *horizontal* and *vertical axes* are both physical properties of the tone. The human part appears in the *curves* themselves. Every point on a single curve describes a sine tone with the same loudness. The *dashed line* corresponds to absolute threshold, the weakest sound the average human can hear

To construct an equal-loudness contour requires an experimenter and a listener. The experiment begins when the experimenter presents a sine tone having a frequency of 1,000 Hz and some chosen level, say 20 dB. The listener listens to that tone. Then the experimenter changes the frequency, for instance to 125 Hz, and asks the listener to adjust the level of that 125-Hz tone so that it is just as loud as loud as the 1,000-Hz tone.

In this example, where the sound level of the 1,000-Hz sine is 20 dB, it is found that the sound level of the 125-Hz sine must be adjusted to 33 dB for equal loudness. These two points (1,000 Hz, 20 dB) and (125 Hz, 33 dB) are on the same equal-loudness contour, and, by definition, this contour is called the 20-phon loudness level. *The frequency of 1,000 Hz is always the reference frequency for the phon scale.* In what follows we will refer to a loudness level in phons by the Greek symbol Φ (Phi, for physical). The level called 0-dB in Fig. 12.1 corresponds to the nominal threshold, 10^{-12} W/m^2.

12.2 Loudness

The loudness level, expressed in phons, Φ, is a frequency-compensated decibel scale, but it is not yet a measure of loudness. A true measure of loudness would scale with the personal sensation of magnitude, but there is no reason to suppose that the numbers 10 phons, 20 phons, etc. will do that. After all, the scale of phons is tied to the decibel scale at 1,000 Hz, and the decibel scale was invented without much consideration of human perception of loudness. A logarithmic scale, like the decibel scale, is convenient mathematically, but careful experiments on human listeners have concluded that loudness is not a logarithmic function of intensity. Loudness does not follow the decibel scale well. Instead, many experiments done over the past 75 years have shown that the rule for loudness is a different kind of function. It is a exponential law, as follows:

$$\Psi \propto 10^{0.03\Phi}, \tag{12.1}$$

where the symbol Greek Ψ (Psi for psychological) stands for loudness. An alternative way to write Eq. (12.1) is as

$$\Psi \propto 1.0715193^{\Phi}. \tag{12.2}$$

These proportionalities don't enable you to calculate any absolute numbers, but they do enable you to calculate ratios. Suppose there are two tones, numbered "1" and "2", that have loudness levels of Φ_1 phons and Φ_2 phons. Then the ratio of the loudnesses is given by

$$\Psi_2/\Psi_1 = 10^{0.03(\Phi_2-\Phi_1)} \tag{12.3}$$

The formula in Eq. (12.3) follows immediately from the proportionality in Eq. (12.1). Ratios can be useful concepts, as shown in the following example.

Example:

Question: There are two sine tones, "1" and "2." These tones have different frequencies and tone 2 is 10 phons greater than tone 1. How much louder is tone 2, compared with tone 1?

Answer: Although the frequencies are different, the fact that loudness levels are given in phons makes it possible to use Eq. (12.3). Plugging into Eq. (12.3) we find:

$$\Psi_2/\Psi_1 = 10^{0.03\cdot(10)} \tag{12.4}$$

or

$$\Psi_2/\Psi_1 = 10^{0.3} \tag{12.5}$$

From the inverse log function on a calculator we find that $\Psi_2/\Psi_1 = 1.995$. That's close to 2. Tone 2 is about two times louder than tone 1.

The example above has illustrated the convenient rule:

Rule 1: *For every increase of 10 phons, the psychological sensation of loudness doubles.*

Loudness vs Sound Level Because we regularly make physical measurements of the sound intensity in terms of the sound level in decibels, it would be really convenient to be able to relate loudness to the sound level, L. There are two requirements that need to be fulfilled in order for us to be able to do that. First, we need to consider a single frequency at a time. Thus we can compare the loudnesses of two tones that have different levels but the same frequency. In other words, we can ask about the effect of turning up the volume control. The second requirement is that the tones involved need to have particular levels and frequencies such that the equal-loudness contours are approximately parallel. A glance at the contours of Fig. 12.1 shows that this second requirement is met by many everyday sounds, having frequencies between 250 and 8,000 Hz and levels greater than 40 dB.

As a practical matter, therefore, we can rewrite Eq. (12.1) as

$$\Psi \propto 10^{0.03L} \tag{12.6}$$

We expect that this relationship will fail at frequencies below 125 Hz, where the equal loudness contours are not parallel but are squished together. We expect that this relationship will hold good in many practical circumstances. Exercises at the end of the chapter will assume that it holds good.

Given the relationship in Eq. (12.6) it follows that if there are two tones of the same frequency with levels L_1 and L_2, then the loudnesses are related by the ratio

$$\Psi_2/\Psi_1 = 10^{0.03(L_2-L_1)} \tag{12.7}$$

A special case of Eq. (12.7) leads to the convenient rule:

Rule 2: *For every increase of 10 dB, the psychological sensation of the loudness of a tone doubles.*

Rule 2 is more convenient than Rule 1 because it uses decibels instead of phons. But Rule 2 is not as general as Rule 1 because it requires that special conditions hold—constant frequency and parallel equal-loudness contours.

Loudness vs Intensity From Chap. 10, we know that the sound level (L) in decibels can be related to the sound intensity (I) expressed in watts per square meter. The relationship is that $I \propto 10^{L/10}$. For tones in the practical region of frequency and level, it follows from Eq. (12.6) that

$$\Psi \propto I^{0.3}. \tag{12.8}$$

This relationship is a power law. It says that loudness grows as the 0.3 power of the intensity. This statement is close to saying that the loudness grows as the cube root of the intensity. The cube root would be the 1/3 (or 0.333...) power, and that is close to 0.3.

Fig. 12.2 The 0.3 power law is a compressive function. As the input goes from 1 to 10, the output goes from 1 to 2

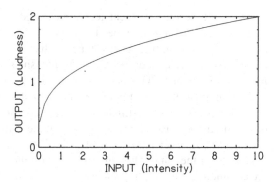

This is quite an amazing rule. Suppose you have a tone with a sound level of 50 dB. Therefore, the intensity is 10^{-7} W/m². If you increase the level to 60 dB the intensity becomes 10^{-6} W/m², and the loudness becomes twice as great. If you increase the level by another 10 dB the intensity becomes 10^{-5} W/m², and again the loudness doubles. This behavior indicates a highly compressive function. Overall, starting from 50 dB, the intensity has increased 100 times but the end result is only four times louder according to human ears. This behavior is called, "compressive" because the output grows slowly as the input increases. Here, one imagines the input to be the physical intensity, put into the human auditory system, and the output is the listener's sensation of loudness. A graph of a compressive function has a shape like Fig. 12.2.

Example:

Question: There are two sine tones, "1" and "2." These tones have the same frequency, but tone 2 is 300 times more intense than tone 1. How much louder is tone 2, compared with tone 1?

Answer: From the proportionality in Eq. (12.8), you know that

$$\Psi_2/\Psi_1 = (I_2/I_1)^{0.3} \tag{12.9}$$

We are told that $I_2/I_1 = 300$, and so $\Psi_2/\Psi_1 = (300)^{0.3} = 5.5$. Thus tone 2 is about five times louder than tone 1.

12.3 Psychophysics

The delicate matter of scaling a psychological magnitude such as loudness is the subject of the science of psychophysics. The psychophysicist tries to determine the relationship between physical magnitudes, measured with ordinary electronic instrumentation, and psychological magnitudes as reported by human (or animal) observers.

There are several different techniques used to determine psychological magnitudes. In a *direct* technique, the experimenter just asks subjects to rate their sensations of magnitude. For instance, a listener might be presented with 1,000-Hz sine tones having a dozen different intensities. The listener is asked to rate the loudness of the tones. This is an example of a *magnitude estimation* task.

It is not clear that this kind of experiment should work. As you can imagine, if you ask ten listeners to estimate the magnitude of a tone, you are likely to get ten different answers. However, if the magnitude estimation experiment is continued for many hours over the course of a week or two, certain clear patterns begin to emerge from the data. Although different listeners may use different scales, there are common features (ratios mostly) in their responses.

After the experimenter has found stable numerical responses to tones of different intensities in the estimation task, the experiment is reversed to form a *magnitude production* task. This time, the experimenter gives the listener the numbers, and the listener adjusts a volume control to vary the intensity of the sound. If the listener previously gave numbers on a scale of 1–100 in the estimation task, then the experimenter will give the listener the same range of numbers for the production task. After another week of experimenting, the data from the two tasks are averaged, and the result is a function showing the relationship between physical magnitude and psychological magnitude. For the loudness experiment, the relationship is that given in Eq. (12.1).

The Sone Scale   The exponent of 0.3, as in Eq. (12.8), is the basis for the current national and international loudness scale known as the *sone* scale. The reference for this absolute scale is that a 1,000-Hz sine tone with a level of 40 dB SPL shall have a loudness of one sone. Therefore,

$$\Psi(\text{sones}) = \frac{1}{15.849}\left(\frac{I}{I_0}\right)^{0.3}, \tag{12.10}$$

or

$$\log \Psi(\text{sones}) = -\log(15.849) + 0.03 \times 10\log\frac{I}{I_0}. \tag{12.11}$$

It follows by definition that any tone with a loudness level of 40 phons has a loudness of one sone, and sones can be calculated from phons by the equation

$$\log \Psi(\text{sones}) = -1.2 + 0.03 L_\varphi, \tag{12.12}$$

where L_φ is the loudness level in phons.

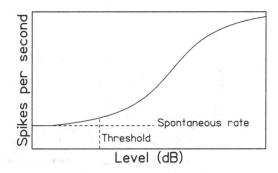

Fig. 12.3 Neural firing rate vs sound level in dB. The threshold is defined as the level where the firing rate becomes 10% greater than the spontaneous rate

12.4   Neural Firing Rate

A scientist finds it natural to try to understand the perception of loudness from what is known about the auditory system and the properties of neurons that process sounds in our brains. It is very compelling to imagine that the loudness sensation is related in some way to the total number of neural spikes transmitted to the central auditory system by the auditory nerve.

As the intensity of a sound increases, the firing rate of neurons in the auditory nerve also increases, but the whole story is not as simple as that. First, even when there is no sound at all, the neurons fire occasionally. This spontaneous firing rate may be about 50 spikes per second. Now suppose that a very weak tone is turned on. For very low intensity levels a neuron's firing rate remains at the spontaneous rate, as shown in Fig. 12.3. As the intensity increases it eventually passes a threshold where the neural firing rate starts to become greater than the spontaneous rate. As the intensity continues to increase, through a range of 30 or 40 dB above threshold, the neuron fires more and more often. But eventually, as the intensity increases still further, the neuron starts to saturate. Its firing rate no longer goes up rapidly with increasing intensity. In other words, neural firing rate looks like a compressive function of intensity. It is a good bet that the loudness of a sound is closely related to the number of neural spikes received by the central auditory system in a 1-s interval, i.e. to the neural firing rate. Therefore, it seems likely that the compressive behavior of loudness observed in the psychology of human hearing is closely related to the compressive behavior of neural firing rate observed physiologically.

12.5   Excitation Patterns

We must remember that the neurons described above in the discussion of neural firing rate are originally excited by a vibration pattern (excitation pattern) on the basilar membrane. The neurons that are active are those at places on the basilar membrane that are excited by frequencies in the tone. If the tone is a sine tone, there

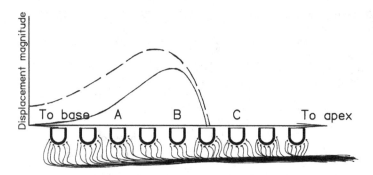

Fig. 12.4 Cartoon showing basilar membrane motion for a sine tone (*solid*) and a more intense sine tone (*dashed*). Three representative hair cells are given labels A, B, and C

is only one frequency and the excitation pattern has a single peak, as shown by the solid line in Fig. 12.4. Therefore, only a few neurons are excited. As the intensity grows, the excitation pattern becomes somewhat wider, as shown by the dashed line, and additional neurons are excited. This is a second mechanism for transmitting the sensation of loudness to the brain. Not only do active neurons become more active (e.g., the neuron at location B) but additional neurons (e.g., the neuron at location A) start to fire beyond their spontaneous rate. The additional spikes contributed by the neuron at A will help somewhat to generate the loudness sensation when the neurons near location B saturate. However, because of neural compression, a law of diminishing returns applies when the only neurons that contribute are those that respond to a single frequency. To increase the total firing rate a lot, to get a really loud sound, you need more than one frequency. You need a complex tone. For instance, adding a low-frequency component should excite the neuron at location C. A broadband tone with components at frequencies that span the audible range excites all places on the basilar membrane and should be particularly loud. The compressive character of neural firing means that the best way to generate loudness is to spread the available power over a broad frequency band rather than over a narrow band where it excites only a few neurons. An example of this effect appears in the next section.

12.6 Complex Sounds

To review where we have been in this chapter, recall that the chapter began with sine tones with particular levels (L in dB) and frequencies (f) and put these tones on equal-loudness contours denoted by loudness level (Φ in phons). Next, we related the loudness level (Φ) to the sensation of loudness (Ψ) through an exponential law. Finally we identified a practical region of levels and frequencies where the equal-loudness contours were approximately parallel, and we related loudness (Ψ) to level (L) and then to intensity (I). We found that loudness was proportional to the 0.3 power of the intensity. We related this power law to facts about the auditory nervous system.

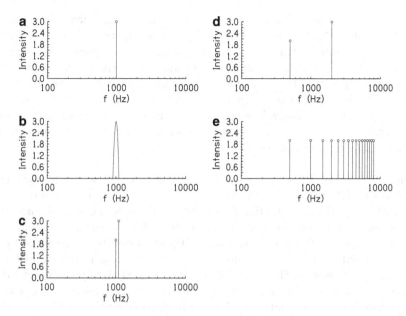

Fig. 12.5 Spectra for loudness calculations: (**a**) sine, (**b**) whistle, (**c**) two closely-spaced sines, (**d**) two separated sines, and (**e**) complex periodic tone. The *horizontal axis* is a logarithmic frequency scale, an approximation to the scale of the real basilar membrane

After all this work, it is a little disappointing to realize that the only sounds that we have considered so far are sine tones (Fig. 12.5a). The real world consists of complex sounds, made up of many sine tone components. Fortunately, it is not too hard to use what we already know to make good guesses about the loudness of complex sounds. The key to our approach is to assume that the sensation of loudness ought to depend on the total number of neural spikes received by the brain. We consider three kinds of sounds in turn.

1. Narrow-band sounds: If you whistle, you create a tone that is almost a sine tone, but it fluctuates in time. This fluctuation appears spectrally as a narrow band of noise, centered on a frequency that corresponds to the pitch of the whistled tone (Fig. 12.5b). The neurons that respond to the narrow band of noise are essentially the same neurons that would respond to a sine tone and we feel confident about using the entire sine-tone model to calculate the loudness of a narrow-band noise. All the components that make the tone noisy are compressed in the same way.

A second example of narrow-band sounds is a few (for instance two) sine tones with frequencies that are close together, e.g., 1,000 and 1,100 Hz (Fig. 12.5c). These tones excite mostly the same neurons and the excitations from both tones are compressed together. Thus if one tone has an intensity of 2 units and the other tone has an intensity of 3 units, the loudness will be proportional to $(2+3)^{0.3} = 5^{0.3}$ or 1.62.

2. Separated sine tones: If there are two sine tones with quite different frequencies then two quite different regions of the basilar membrane will be active and quite different sets of neurons will be excited. For instance, the frequencies might be 500 and 2,000 Hz (Fig. 12.5d). Excitations from these two sets of neurons will be compressed separately. Thus if one tone has an intensity of 2 units and the other tone has an intensity of 3 units, the loudness will be proportional to $(2)^{0.3} + (3)^{0.3} = 1.23 + 1.39 = 2.62$. You will notice that the loudness of the separated tones is considerably greater than the loudness of tones that are close together in a narrow band, i.e., 2.62 is greater than 1.62.

3. Harmonic tones: Let's consider a periodic complex tone with a fundamental frequency of 500 Hz and 16 harmonics (Fig. 12.5e). This tone has components at 500, 1,000, 1,500, ..., 8,000 Hz. To find the loudness of this tone we need to add up the contributions from all 16 components. What is interesting about this tone is that the low-frequency components 500 and 1,000 Hz are separated by a factor of 2 and excite different neurons in the auditory system. But the highest-frequency components, 7,500 and 8,000 Hz, are separated by only a factor of 1.07, and they excite mostly the same neurons. Thus, both kinds of addition—separated components and narrow band—must be done for such a sound.

12.7 Critical Band

The loudness calculations above point up the need for a way to decide whether the spectral components of a sound will excite the same neurons in the auditory system or different neurons. Strong compression occurs when only the same neurons are involved. Thoughts like these led to the concept of the *critical band*. If two spectral components are separated in frequency by more than a critical band, they excite different neurons. Two components that are separated by less than a critical band excite the same neurons. It should be evident that the all-or-nothing character of this definition of the critical band is only an approximation to what is really a continuum of interaction along the basilar membrane and in higher auditory centers. Nevertheless, psychoacoustical experiments have found reasonable reproducibility in measurements of the critical band width.

One way to measure the width of a critical band is by loudness, as suggested above. As the frequency separation of two tones increases, the loudness starts to increase when the separation exceeds a critical band width. Another way to measure is by *masking*. A weak tone is masked by an intense band of noise if the frequency of the tone and the frequencies of the noise fall within the same critical band. Masking means that the weak tone cannot be heard because of the intense noise. It is an amazing fact that if the weak tone and the intense noise band are separated in frequency by more than a critical band then there is almost no masking at all!

Fig. 12.6 Critical band widths are shown by the *solid line*. They were measured using bands of masking noise having a notch with variable width. One-third octave widths are shown by the *dashed line*. Because the plot is a log–log plot, the one-third octave reference is a *straight line*

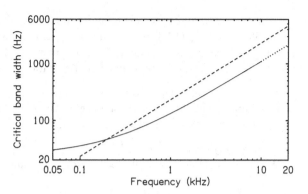

Critical band widths determined from masking experiments are given in Fig. 12.6. The dashed line is a reference showing bandwidths that are one-third of an octave wide. Evidently the critical band is wider than 1/3 octave at low frequencies and narrower than 1/3 octave at high frequencies.

Exercises

Exercise 1, Equal loudness contours
    What are the axes of the equal loudness contours?

Exercise 2, Phons
    What is the loudness level in phons of a 300-Hz tone having a sound level of 40 dB?

Exercise 3, Sub-threshold or supra-threshold
    Is it possible to hear a 200-Hz tone with a sound level of 10 dB?

Exercise 4, Threshold
    In terms of the equal loudness contours, why is the threshold of hearing said to be 0 dB?

Exercise 5, Following the curves
    What is the level of a 4,000-Hz tone that is as loud as a 125-Hz tone having a sound level of 50 dB?

Exercise 6, How to make loudness
    Here is a loudness competition. You start with a 1,000-Hz sine tone having an intensity of 10^{-6} W/m². To increase the loudness, you are allowed to add another sine tone, also with an intensity of 10^{-6} W/m². Your goal is to choose the frequency of that added tone in order to lead to the greatest possible loudness. Which frequency do you choose? (a) 100 Hz, (b) 1,000 Hz, (c) 1,100 Hz, (d) 6,000 Hz.

Exercise 7, The compressive law of loudness

(a) Show that Eq. (12.8) agrees with the idea that increasing the intensity by a factor of 10 doubles the loudness. (b) If the intensity of tone 2 is 50 times greater than the intensity of tone 1, how much louder is tone 2 compared to tone 1?

Exercise 8, The compressive law used in reverse

How much must the intensity of a tone be increased in order to create (a) four times the loudness; (b) 16 times the loudness?

Exercise 9, Why is there a valley?

Why are equal loudness contours lowest in the frequency range from 3 to 4 kHz?

Exercise 10, Loudness and dB

The band conductor asks the trombone player to play louder. The trombone player increases the level by 7 dB. How much louder is the sound?

Exercise 11, Listening to your car radio

You are driving at high-speed on a highway listening to your car radio. There is a lot of road noise, and you need to turn up the level of your radio to hear the music. When you exit the highway and come to a stop at the traffic signal, your radio sounds much too loud. How does this relate to neural firing? It should be evident that the total neural firing rate in your auditory system is less when you are stopped than when you are driving at high speed. How can you explain the fact that the radio is louder when you are stopped?

Exercise 12, Terminology

Bozo objects to the last sentence of the exercise immediately above. He says that the radio isn't really louder when you are stopped—it just *sounds* louder. Set Bozo straight on the scientific definition of loudness.

Exercise 13, Bathroom fans

Consumers Reports (January 2004) rates the noise of bathroom fans in units of sones. Quiet fans are rated from 0.5 to 1.2 sones, and the loudest are rated at 4 sones. How much louder is a 4-sone fan compared to a 0.5-sone fan?

Exercise 14, Contour maps

As you look at a contour map, you see that at some places on the map the contours are far apart and at other places the contours are close together. From the spacing of the contours, you learn whether the terrain is rather flat or whether it is steeply rising. What can you infer from the equal loudness contours in Fig. 12.1. Compare low-level signals at 62 Hz with low-level signals at 1,000 Hz.

♠

Chapter 13
Pitch

Pitch is the psychological sensation of the highness or the lowness of a tone. Pitch is the basis of melody in music and of emotion in speech. Without pitch, music would consist only of rhythm and loudness. Without pitch, speech would be monotonic—robotic. As human beings, we have astonishingly keen perception of pitch. The principal physical correlate of the psychological sensation of pitch is the physical property of frequency, and our keen perception of pitch allows us to make fine discriminations along a frequency scale. Between 100 and 10,000 Hz we can discriminate more than 2,000 different frequencies!

13.1 Pitch of Sine Tones: Place Theory

Because frequency is so important to pitch, we begin the study of pitch with a sine tone, which has only a single frequency. From Chap. 11 you know that the basilar membrane is selective for frequency. Different frequencies cause the membrane to vibrate maximally at different locations. This, in turn, causes particular hair cells to be maximally excited by particular frequencies. Each hair cell is connected to particular neurons in the auditory nerve, and this establishes a neural selectivity for frequency that is maintained throughout the entire auditory system. The frequency selectivity of neurons is called "tonotopic analysis," *tono-* for tone frequency, and *topic* for location, meaning that different frequencies excite slightly different locations in the brain.

Nothing could be more natural than to assume that the sensation of pitch is the result of tonotopic analysis. This idea is called the "place theory" of pitch perception. According to this theory, you hear different pitches because different neurons (different places in your brain) are excited. From years of research, including recent studies of persons with cochlear implants, we know that this place theory is capable of accounting for many aspects of pitch perception. However, there is more...

W.M. Hartmann, *Principles of Musical Acoustics*, Undergraduate Lecture Notes in Physics, DOI 10.1007/978-1-4614-6786-1_13,
© Springer Science+Business Media New York 2013

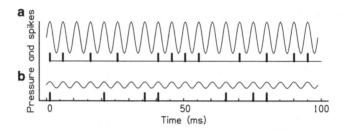

Fig. 13.1 The intense tone in (**a**) produces more neural spikes than the weak tone in (**b**), but both tones produce synchronized spikes

13.2 Pitch: Timing Theory

You know about neural firing rate. You know that a greater firing rate corresponds to greater loudness. You know that a high firing rate in certain neurons encodes pitch—the place theory mechanism. Now you need to recall a third property of the neural code, introduced in Chap. 11, namely *neural synchrony*. The firing of a hair cell and the spikes in the neurons connected to it synchronize to the input tone. This happens because a hair cell tends to fire on a particular phase of the pressure wave in the cochlea. Figure 13.1 shows the firing of a hair cell that fires on waveform peaks. Part **a** shows 1/10 s of a 200-Hz sine tone of moderate intensity. This tone causes 12 neural spikes, so the firing rate is about 120 spikes per second, but the spikes don't come at random times. The time intervals between spikes are integer multiples of a period, specifically: 5, 10, 10, 15, 5, 5, 5, 15, 10, 10, 5 ms. This sequence is only an example. In the next 1/10 s, the sequence of intervals will no doubt be different, but all the intervals will be multiples of 5 ms.

In this way, the timing of the neural spikes encodes the period of the tone, and this provides another cue to the brain for the pitch of the tone. The idea that the synchrony of neural spikes leads to the sensation of pitch is called the "timing theory" of pitch perception. Part **b** of Fig. 13.1 shows another 200-Hz tone, but less intense. There are only seven spikes in 1/10 s, but they too are synchronized with the signal.

So far as the timing theory is concerned, the intervals are still multiples of 5 ms and the timing information about pitch is not different for this weaker tone.

13.3 Pitch of a Complex Tone

Most sounds that we encounter are complex. They have multiple frequency components. If the tone is periodic then the components are harmonic—their frequencies are integer multiples of a fundamental frequency. Figure 13.2 shows a very typical tone. It could be a musical instrument tone or a sustained spoken vowel. Ten

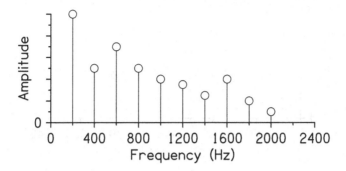

Fig. 13.2 A complex tone with ten harmonics of 200 Hz

harmonics are shown. There are more, but they are normally weaker than the first half dozen. The fundamental is the strongest harmonic, which is typical, but for many other complex tones the fundamental is not the strongest. In fact, the fundamental might not even exist. Nothing that we say about pitch requires that the fundamental be strong or even that it be present.

The tone in Fig. 13.2 has a pitch of 200 Hz. That means that if you listen to that tone for about a second or less, and try to sing what you heard, you will sing a tone with a fundamental frequency of 200 Hz. Or, to be more precise, you might have a sine-tone generator with a knob that controls the frequency. When you adjust the sine frequency to match the pitch of the complex tone it turns out that you set the generator to 200 Hz.

This experimental result makes the timing theory look very good. Because the components are all harmonics of 200 Hz, the period of the waveform is 1/200 s. The prediction of the timing theory and the experimental results agree that (1) the pitch depends on the period and does not depend much on the relative strengths of the harmonics. (2) The pitch does not depend on the fundamental or other low harmonics being present in the spectrum. For instance, a tone with the spectrum of Fig. 13.3 also has a period of 1/200 s and it also leads to a pitch of 200 Hz. (3) The pitch is insensitive to the overall intensity of the tone. By contrast, it's not clear what the place theory has to say about a complex tone. The place theory might predict that a tone with ten harmonics, like Fig. 13.2, should lead to ten different pitches. After all, the ten different frequencies in the tone excite the basilar membrane at ten different places.

Before giving up on the place theory though, we need to do some more experimenting. Suppose we have a spectrum like Fig. 13.4 with a 200-Hz fundamental component and a 10th harmonic component at 2,000 Hz and no other components. Once again, the period is 1/200 s, but this time one clearly hears two distinct pitches, one at 200 Hz and the other at 2,000. This observation agrees with place theory and not with timing theory. The obvious conclusion is that neither the place theory of pitch perception nor the timing theory of pitch perception is good enough to account for the sensations produced by all possible tones. We shall continue to try to find a better theory.

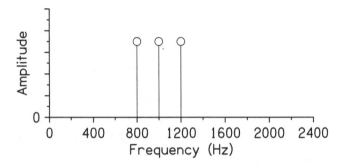

Fig. 13.3 The spectrum has components at 800, 1,000, and 1,200 Hz, corresponding to the 4th, 5th, and 6th harmonics of 200 Hz. According to most listeners, the pitch is 200 Hz

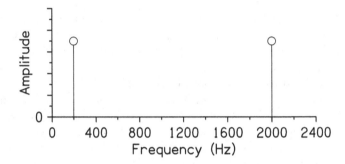

Fig. 13.4 The spectrum has components at 200 and 2,000 Hz, corresponding to the first and 10th harmonics of 200 Hz. The two components are heard separately—analytic listening

13.4   The Template Theory

The template theory (or template model) of pitch perception deals with some of the problems of the place theory. The template theory says that from early experience (maybe even prenatal!) you have learned to associate a tone having a harmonic spectrum with a pitch corresponding to the fundamental frequency (or period) of the tone. When you hear a tone, you subconsciously match its spectrum with a template (a spectral pattern) stored in your memory. The pattern that best matches the harmonics in the tone is the pattern that gives the sense of pitch.

An advantage of the template theory is its flexibility. For instance, it can deal successfully with slightly inharmonic tones. Suppose one starts with the 200-Hz complex tone of Fig. 13.3 and shifts each component frequency by 30 Hz. The components then have frequencies 830, 1,030, and 1,230 Hz. This tone does not have harmonic components, and it is not strictly periodic (Fig. 13.5).

Experiments with this tone show that its pitch is not as well defined as the pitch of a periodic tone, but that listeners rather consistently find the pitch to be about 206 Hz. This result provides support for the template theory. According to the

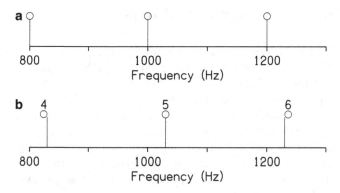

Fig. 13.5 (**a**) The spectrum from Fig. 13.3 is replotted here on a different scale. The fundamental frequency and pitch are 200 Hz. (**b**) The spectrum, shown by *lines*, has components at 830, 1,030, and 1,230 Hz, and these are not harmonics. However, the spectrum fits reasonably well with a harmonic template having a fundamental frequency of 206 Hz and harmonic numbers 4, 5, and 6, shown by *circles*

template model, the listener chooses a pitch for which the harmonics of a harmonic template would best agree with the components that are actually present. In this case, a 206-Hz template would have a fourth harmonic at $4 \times 206 = 824$ Hz, a fifth harmonic at $5 \times 206 = 1,030$ Hz, and a sixth harmonic at $6 \times 206 = 1,236$ Hz. Compared to the actual components, the template fourth harmonic would be too low $(824 < 830)$, the template sixth harmonic would be too high $(1,236 > 1,230)$, and the template fifth harmonic would be exactly right. One can make the case that if one insists on a harmonic template then 206 Hz leads to the best match to the three components actually present. According to the template theory, that is why the brain perceives a pitch of 206 Hz.

Template Model Calculations

To calculate a pitch from a set of spectral frequencies using the template model, you imagine that each frequency in the spectrum suggests a fundamental. You then average all those suggestions. To begin this process, you first need to decide what harmonics shall correspond to the frequencies that you are given.

For instance, given spectral components at 830, 1,030, and 1,230 Hz you might decide that these look like harmonics 4, 5, and 6. Therefore, the suggested fundamentals are 830/4, 1030/5, and 1230/6. The average of these is

$$\frac{1}{3}(830/4 + 1030/5 + 1230/6) = 206 \text{ Hz,}$$

in agreement with the more intuitive calculation above. The choice of harmonics 4, 5, and 6 was consistent with the general observation that listeners prefer templates with consecutive harmonics.

13.5   Pitch as an Interpretative Percept

Attempts to build a purely mechanistic theory for pitch perception, like the place theory or the timing theory, frequently encounter problems that point up the advantages of less mechanistic theories, like the template theory. Often, pitch seems to depend on the listener's interpretation. Although the auditory system segregates many of the harmonics of a complex tone like Fig. 13.2 into separate neural channels, the brain finds little benefit in hearing out the different harmonics and tends to hear the tone as a single entity with a single pitch. However, if the spectrum has an unusually large gap, like the gap between harmonics 1 and 10 in Fig. 13.4, the brain decides that there are two entities and assigns two pitches. In this case the place model works.

But the place model does not explain why it is so important for our perception that the partials of a tone should be harmonic. A harmonic tone holds together and is perceived as a single entity with a single pitch. Inharmonic tones, by contrast, often separate into several entities each with a different pitch.

For an inharmonic tone, the brain might hear the complex as a single tone with a pitch for which the harmonic template would best fit the inharmonic components, or it might hear out the individual components themselves. Hearing a single tone is known as *synthetic listening* because the listener synthesizes a single pitch from the components. Hearing the individual components is known as *analytic listening* because the listener perceives the spectral analysis originally performed in the cochlea. Listening to a tone for a long time may cause a listener to hear out individual harmonics—analytic listening. That is why the discussion of the tone in Fig. 13.2 suggested listening for a second or less in order to hear a synthetic pitch. Experiments with ambiguous tones (e.g., Exercise 5 below) show that some listeners tend to be synthetic and others tend to be analytic. The variety of experience in pitch perception seems to preclude any really simple mechanistic model. However, that does not prevent us from continued efforts to find one.

Visual Templates

The interpretative nature of pitch perception has visual analogues, for example Fig. 13.6. You can probably recognize a word in the symbols of that figure, and yet you have never seen the word written like that before at any time in your life. How did you do it? The answer is that the symbols are fairly close to letters for which you have templates stored in your brain, and the template enables you to make the association between the image on the paper and the letter. Also, the word itself is familiar. As you learned to read, you acquired the habit of linking symbols to form a word as a single entity. On the other hand, the last symbol may remind you of an umbrella. Unusual letters can separate like that into several entities, each with its own identity. It's a matter of interpretation, and your brain is engaged in that activity full time, both in the visual world and in the auditory world.

$$\{ \Pi \mathsf{J}$$

Fig. **13.6** Three (unfamiliar?) symbols can be interpreted as a word or segregated individually

13.6 Absolute Pitch

Absolute pitch (AP), sometimes called "perfect pitch," refers to the ability to name musical notes without any reference tone. A person with absolute pitch can immediately identify a note played on a piano. Also, that person can sing a note on demand. For instance, if you ask an AP possessor to sing a "D," that person will sing a note in a comfortable octave that is much closer to "D" than to any other note of the scale. Absolute pitch is different from relative pitch (RP) in which a listener can identify a note when given a reference. For instance, an RP possessor can sing a "D" if he or she knows that the note just played on a piano is the note called "G."

Absolute pitch in hearing can be compared with absolute color in vision. Everyone with normal color vision has a sense of absolute color in that we can immediately name a color. For AP possessors, musical pitches are identifiable like colors are identifiable for the rest of us.

Less than 1% of the population has AP, and it does not seem possible for adults to learn AP. By contrast, most people with musical skills have RP, and RP can be learned at any time in life. AP is qualitatively different from RP. Because AP tends to run in families, especially musical families, it used to be thought that AP is an inherited characteristic. Most of the modern research, however, indicates that AP is an acquired characteristic, but that it can only be acquired during a brief critical interval in one's life—a phenomenon known as "imprinting." Ages 5–6 seem to be the most important.

The musical family tendency can be understood in terms of imprinting. If a child hears songs haphazardly sung in different keys on different days or by different people, the child learns that what is important about tunes is the relative pitches of the tones. If a child listens to a family member repeatedly practicing a piece of music—always in the same key—or if the child regularly practices, the child may learn that what is important is the absolute pitch of tones.

Exercises

Exercise 1, Tonotopic separation

Go back to Chap. 11 and look at the frequency map according to Greenwood showing the place of maximum stimulation on the basilar membrane vs frequency.

Fig. 13.7 Spectrograms
showing the tone pairs in
Experiments **a** and **b** called
"Analytic or synthetic pitch"

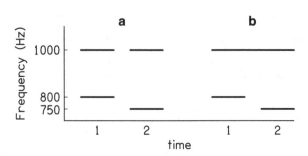

How far apart are places 200 and 2,000 Hz? This large distance is probably the
reason that a tone with only these components is not perceptually fused into a single
200-Hz tone. Instead the individual components stand out.

Exercise 2, Neural synchrony: interspike interval histogram

Figure 13.1 Part **a** shows neural spikes in response to an intense 200-Hz sine tone.
What time intervals occur between neural spikes? Make a histogram plot that shows
the number of spikes separated by one cycle, by two cycles, and by three cycles.
When you are done, you will have made an interspike interval histogram—a very
useful representation used everyday by neuroscientists. The only difference is that
they have thousands of intervals in their data, and you have only 11. Make another
histogram for the weak tone in Part **b** of Fig. 13.1.

Exercise 3, A case of the missing fundamental

A complex tone has three components with equal amplitudes and frequencies:
750, 900, and 1,050 Hz. From the template model, what pitch do you expect?

Exercise 4, The template model

(a) Complex tone A has three components with equal amplitudes and frequencies: 420, 620, and 820 Hz. (b) Complex tone B has five components with equal
amplitudes and frequencies: 420, 620, 820, 1,020, and 1,220 Hz. From the template
model, what pitches do you expect?

Exercise 5, Analytic or synthetic pitch

There is a sequence of two tones, first *1* then *2* as shown in Fig. 13.7.

Tone *1* has two equal-amplitude components at 800 and 1,000 Hz.

Tone *2* has two equal-amplitude components at 750 and 1,000 Hz.

Do you predict that the pitch goes up or down in Experiment **a**? How about
Experiment **b**?

♠

Chapter 14
Localization of Sound

The ability to localize the source of a sound is important to the survival of human beings and other animals. Although we regard sound localization as a common, natural ability, it is actually rather complicated. It involves a number of different physical, psychological, and physiological, processes. The processes are different depending on where the sound happens to be with respect to the your head. We begin with sound localization in the horizontal plane.

14.1 Horizontal Plane

The horizontal plane is the plane parallel to the ground that includes your two ears and your nose. (It is a fact of geometry that it takes three points to define a plane.) We regard your head as a center point and think about sound sources located at other points in this plane. Those other points can differ in distance and azimuth with respect to your head. We begin with azimuth. The azimuth is an angle typically measured in degrees from the forward direction. Therefore, a point directly in front of your nose is at $0°$ azimuth. Points in space that are directly to your right and left have azimuths that are $+90°$ and $-90°$, respectively, as shown in Fig. 14.1.

Modern approaches to sound localization in the horizontal plane are restricted to sources that lie in a within the $180°$ range from your extreme left ($-90°$) to your extreme right ($+90°$). The rest of the horizontal plane is behind you, and that becomes a matter to be discussed in connection with vertical plane localization. What is significant about this restricted range of locations in the horizontal plane is that different angles lead to signals that are uniquely different in your left and right ears (unless the source is directly in front). Differences between your ears (interaural differences) are processed by part of your nervous system called the "binaural system." The binaural system consists of several processing centers in the brain stem and midbrain that receive neural spikes from a processing chain that starts with the hair cells in the cochlea. These specialized binaural centers (olivary

W.M. Hartmann, *Principles of Musical Acoustics*, Undergraduate Lecture Notes in Physics, DOI 10.1007/978-1-4614-6786-1_14,
© Springer Science+Business Media New York 2013

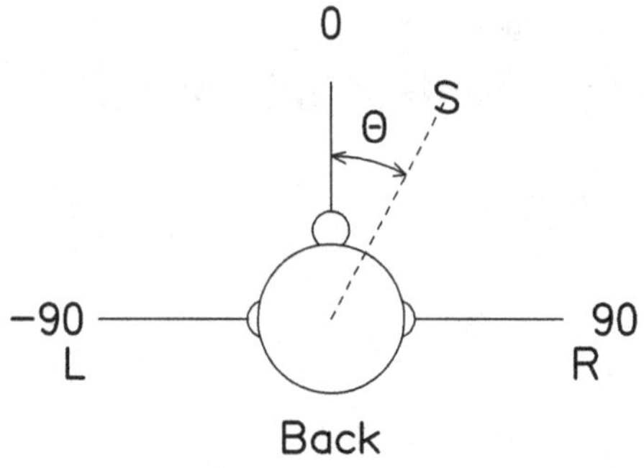

Fig. 14.1 A head and sound
source (symbol S) in the
horizontal plane. The source
has azimuth θ with respect to
the forward direction. A
wavefront from the source
arrives at the right ear before
arriving at the left. Also, the
head partially blocks the
sound wave coming to the left
ear causing it to be less
intense at the left ear

complex, trapezoidal body, and part of the inferior colliculus) are particularly speedy
and efficient. The upper levels of this binaural system are combined with visual
centers that also encode location information. This combination allows the auditory
and visual systems to learn from each other.

There are two main kinds of interaural differences, interaural level differences
(ILDs) and interaural time differences (ITDs). We will deal with these in turn.

14.1.1   Interaural Level Differences

If a sound source is located off to your right, then it is closer to your right ear than
to your left. Furthermore, a sound wave from the right has to go around your head
somehow in order to arrive at your left ear. Thus, there are two effects causing the
sound level to be higher in your right ear than in your left. The first is a distance
effect and the second is a "head-shadow" effect. Both lead to an ILD.

It is not hard to show that unless a sound source is very close to your head the
distance effect is not important. In Exercise 14.1 you will show that for a source
that is 1 m away, the distance effect leads to an ILD no more than 1.4 dB. The effect
is even smaller if the source is further away. But the distance effect is of obvious
importance when the source is very close to your head, for instance when there is a
mosquito in one of your ears.

To deal with the head-shadow effect without the distance effect we assume that
the source is a few meters away. The head-shadow effect leads to ILDs that can
be quite large, but the ILDs depend on frequency. Figure 14.2 shows the ILD for
a source that is off to your right at an azimuth of $10°$, $45°$, or $90°$. The vertical
axis shows how much greater the level is in your right ear compared to your left.
The figure shows that at low frequency there is hardly any effect at all. The ILD
is essentially zero. That is because the low-frequency sound has a wavelength
that is much larger than your head diameter, and the sound diffracts around your
head, almost as though it were not there at all. By contrast, at high frequencies the
wavelength is small, and your head represents a significant obstacle. At 4 kHz and
$90°$ the ILD caused by head-shadow reaches 10 dB.

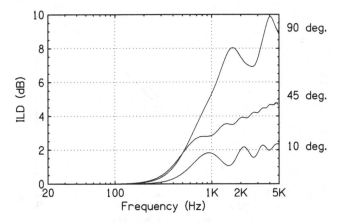

Fig. 14.2 The ILD for a sine tone is shown as a function of the frequency of the tone for three values of the azimuth, 10°, 45°, and 90°. The values were calculated from a equation for sound diffraction around a sphere having the diameter of the average human head. The oscillations in the plots reveal the complexity of the diffraction process. They can lead to some unexpected and confusing effects in controlled experiments

14.1.2 Interaural Time Differences

If a source of sound is located to your right, then every feature of the waveform arrives at your right ear before it arrives at your left ear. The difference in arrival time is known as the ITD. When you first think about it, it is difficult to believe that the ITD could have much effect. On the human scale, sound travels very fast, and the ears are quite close together. It is hard to imagine that the delay of a sound wavefront as it passes first one ear and then the other could be a large enough delay to be perceived. In fact, even for a sound at 90° azimuth (directly to your right) the delay in your left ear is less than 800 μs, equivalent to 0.8 ms. Strange as it may seem, the human binaural system is capable of dealing with interaural differences that small, *and much smaller*. It uses the differences to provide important information about the location of sounds. To understand how ITD localization works there are two things that we have to know. First, we need to know the relationship between the azimuth of the source and the ITD. Second, we need to know about the ability of the binaural system to perceive the ITD.

Source Azimuth and ITD

There is a formula, coming from a simplified theory of sound diffraction, that can be used to calculate the ITD for a source at azimuth θ:

$$\text{ITD} = \frac{3\,r}{v}\sin(\theta), \tag{14.1}$$

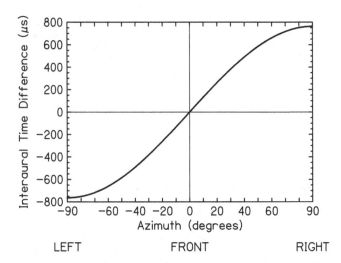

Fig. 14.3 The interaural time difference (ITD) as a function of azimuth from Eq. (14.1). The ITD is zero in the forward direction (azimuth 0°) because a source directly in front is equally distant from the two ears. For a negative value of the azimuth, the ITD is the negative of what it is for the corresponding positive azimuth value

where r is the head radius and v is the speed of sound. A typical human head radius is 8.75 cm, and the factor $3r/v$ is 763 μs. Because the sine function is never greater than 1.0, the ITD can never be greater than 763 μs. The ITD vs azimuth, according to Eq. (14.1), is shown in Fig. 14.3.

The Binaural System and the Usefulness of ITDs The ability of the human brain to make use of the ITD caused by the head is different for different azimuths. The system works best near an azimuth of zero (directly in front), where it is sensitive to an ITD as small as 20 μs, corresponding to a change of 1.5°. At 75° off to the side, the system is less successful. It is only sensitive to an angular difference of about 8°, but that is more precise than visual localization at that large azimuth.

There are important limitations in the ability of the binaural system to make use of ITD. If the frequency of a tone or noise is higher than 1,400 Hz, the waveform fine structure is too fast for the system to follow. Then the binaural system can only register time differences in the envelope (i.e., the amplitude), including the onset. Evidently, the use of ITD depends on the frequency range. To illustrate the different ways that ITD can be used, Fig. 14.4 shows a tone with an abrupt onset and a frequency glide or sweep—musically known as a "glissando."

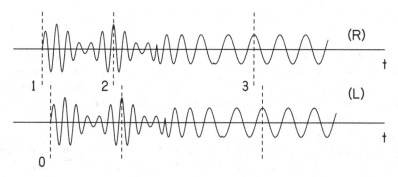

Fig. 14.4 The figure shows a high-frequency tone (1,800 Hz) that quickly sweeps to a low frequency (800 Hz). The high-frequency tone is modulated so its amplitude varies slowly. The source of the tone is 33° to the right. Therefore, from Eq. (14.1), the signal in the right ear leads the signal in the left by 417 μs. The binaural system can respond to this ITD in three ways, shown by numbers between the plots. (1) The onset occurs first in the right ear. (2) The structure in the modulation of the high-frequency waveform occurs first in the right ear. The 1,800-Hz waveform is too fast to observe the ordered structure in the waveform, but the ordered structure in the envelope (amplitude) can be seen. (3) The fine structure in the 800-Hz waveform occurs first in the right ear. Unlike the waveform fine structure at 1,800 Hz, the fine structure at 800 Hz is slow enough for the binaural system to measure

---

Binaural beats

Binaural beats occur when sine tones of slightly different frequency are separately sent to the left and right ears using headphones. Because the two frequencies are not physically combined there are no real beats, but there can be beats in the brain leading to a spatial effect. For instance, given a 500-Hz tone in the left ear and a 501-Hz tone in the right, most listeners hear a tone, with a pitch near 500 Hz, moving back and forth within the head once per second. If the frequencies are increased to 1,000 and 1,001 Hz, the frequencies are high enough that binaural effects are weak and the binaural beats are hard to hear.

To get a visual impression of the binaural effect, return to Fig. 6.4, which was drawn to explain real beats. Imagine that the solid curve represents the signal in your left ear and that the dashed curve represents the signal in your right. Notice that along the time line between 10 and 20 ms, the dashed curve seems to lead the solid curve. Now notice that between 80 and 90 ms the solid curve seems to lead the dashed. Your brain interprets this situation as a source first on the right and then on the left. (Recall that for visual purposes Fig. 6.4 has frequencies of 500 and 510 Hz leading to ten beats per second instead of one beat per second.)

Summary

The sections above have described horizontal (azimuthal) plane localization for steady-state sounds—the continuous parts of sine tones or complex tones with a simple spectral structure and no abrupt onset. Abrupt onsets lead to additional localization information that will be discussed in the section on the *precedence effect*. The interaural differences, ITD and ILD, in the waveform of a steady sound, or a sound with a slow onset, normally work together to localize the sound. The ITD seems to have the stronger influence, but the ITD has the limitation that it is only useful for frequencies below about 1,400 Hz. For tones above the not-particularly-high frequency of 1,400 Hz, the ILD is the main influence on localization, though some additional information is available from the ITD in the modulation of a time-varying signal, as shown in segment (2) of Fig. 14.4.

Sine tones without abrupt onsets have no modulation, as shown by segment (3) in Fig. 14.4. The usefulness of the ITD and the ILD in sine tones has a curious frequency dependence. At very low frequencies (below 200 Hz) the wavelength is so large compared to the size of the head that diffraction causes the levels in the two ears to be almost the same, and the ILD is too small to be useful. The ITD is usable at very low frequencies, though it is not particularly accurate because the period (greater than 5,000 μs) is so long that timing it is difficult. At low frequencies (400–900 Hz) the ITD process becomes highly effective and accurate. At quite high frequencies, above 2,000 Hz, the head provides a reliable shadow on the ear farther from the source, and the ILD is a useful cue. For intermediate frequencies (1,000–2,000 Hz) ITD information is weak or nonexistent. Also, head diffraction becomes a complicated function of both frequency and source azimuth causing the ILD information to be confusing. As a result, sine tone localization is worst in this intermediate frequency region.

14.2 Localization in the Vertical Median Plane

The vertical median plane is the plane that includes the points directly in front of you, directly in back of you, and directly overhead. Points in this plane are symmetrical with respect to your two ears. Therefore, sources in this plane lead to no important interaural differences. Nevertheless, you are able to localize sources in this plane. For instance, you can tell the difference between front and back. One reason that you can do this is that you can turn your head so that the source is somewhat off to the left or right. Then your ability to make left–right localization judgments coupled with your sense of motion enables you to distinguish front from back. However, if the sound is too brief for you to turn your head, or if you are unable to move your head, you can still tell front from back. That's because sounds from the front and back are differently filtered by the anatomy of your head. Your head does not have front–back symmetry (no nose on the back of your head), and this is especially true of your outer ears. The diffraction of short-wavelength sounds from these anatomical features leads to filtering of high-frequency sounds that is different for sounds coming from different directions in the median plane. The most important frequencies are above 6,000 Hz. Your brain has learned to recognize the spectral shapes that these anatomical filters impose on sound waves, and that is what enables you to localize sources in the vertical median plane.

The same kinds of spectral discriminations, in combination with the azimuthal, interaural localization processes, mediate the ability to distinguish between front and back and to perceive the elevation of sources off to the left or right. Experiments on front–back localization show large individual difference. Some people can distinguish directly in front from directly in back with noise bands near 6,000 Hz that are only a few kilohertz wide. Other people require very wide bands extending well above 10 kHz.

14.3   The Precedence Effect

The interaural differences and the spectral cues that are so important to sound localization are determined by the location of the sound source with respect to your head. Perhaps a coin is dropped on the table to your right or there is a talker to your left. The waves that come to your head directly from the source of the sound contain the useful localization cues.

A potential problem occurs when sounds are heard in a room, where the walls and other surfaces in the room lead to reflections. Because each reflection from a surface acts like a new source of sound, the problem of locating a sound in a room has been compared to finding a candle in a dark room where all the walls are entirely covered with mirrors. Sounds come in from all directions and it's not immediately evident which direction is the direction of the original source.

The way that the human brain copes with the problem of reflections is to perform a localization calculation that gives different weight to localization cues that arrive at different times. Great weight is placed on the information in the onset of the sound. This information arrives directly from the source before the reflections have a chance to get to the listener. The direct sound leads to localization cues such as ILD, ITD, and spectral cues that accurately indicate the source position. The brain gives much less weight to the localization cues that arrive later. It has learned that they give unreliable information about the source location. This weighting of localization cues, in favor of the earliest cues, is called the *precedence effect*.

There are a number of rather amazing audio illusions that demonstrate the precedence effect using electronic reverberation simulators. For instance, if one plays a recorded voice from the right loudspeaker of a left–right stereo pair and plays the reverberation of that sound from the left loudspeaker, a listener will hear the sound coming from the right, even if the sound from the left speaker is 8 dB more intense. The precedence effect operates here because the reverberant sound resembles the direct sound but arrives later.

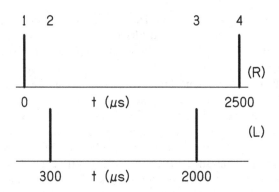

Fig. 14.5 A click pair to study the precedence effect: there are four pressure pulses, two at the right ear and two at the left. Pulses 1 and 2 form a *leading* click. Pulses 3 and 4 form a *lagging* click. However, the pulses come so close together in time that they sound like a *single click image*. (The time axis shows *microseconds*.) The leading click represents a direct sound arriving from the listener's right. The lagging click represents a reflected sound from the listener's left. Even though the reflected sound has an ITD ($-500\,\mu$s) that is larger in magnitude than the direct sound ($300\,\mu$s), the direct sound wins the competition. The listener hears the sound coming from the right because of the precedence effect. The direct sound arrives first

The precedence effect is particularly easy to study with a click pair, as shown in Fig. 14.5. A click has all frequencies, high, medium, and low, and all of the anatomical localization processes are available. Experiments with click pairs have emphasized that the precedence effect relies importantly on high levels of the central nervous system. Infants do not have it—it is developed in childhood. Also, the precedence effect can be affected by context. A long series of click pairs with the leading click on the right, i.e., repeatedly presenting the click pair shown in Fig. 14.5, will bias the system. If that series is followed by a click pair with the leading click on the left, that last click will split into two images—one to the left and one to the right.

14.4 Perceived Auditory Space

Sound localization, as described earlier in this chapter, is an important perceived spatial effect, but it is not the only perceived spatial aspect of our auditory experiences. Other aspects include auditory image *externalization, distance perception,* and *compactness.*

Externalization: The burst of noise when you open a soda bottle is not only localized, it appears to be "out there" somewhere in the space where you live. It is "externalized." Similarly, an electronic burst of noise reproduced by a loudspeaker in a room leads to an externalized auditory image. If you take that same electronic burst of noise and send it to a listener wearing headphones (same noise signal into both left and right headphones), the listener will not perceive an externalized image. Instead, the listener tends to hear the noise within the head.

The distinction between externalized sounds and internalized sounds is important enough that auditory scientists use the term "lateralization," instead of the word "localization," to describe the perceived location of headphone-presented sounds heard within the head. Fortunately for the science of binaural hearing, it happens that there is a rather good correlation between localization and lateralization. Imagine a real sound source—somewhere out there—that produces interaural differences that cause the image to be localized half way between the front and the side. If the same interaural differences are produced using headphones, the image will be lateralized about half way between the center of the head and the extreme right. Therefore, much of what we know about sound localization was actually discovered in headphone experiments studying lateralization. Nevertheless, externalization is a separate perceptual phenomena, and for many years it was something of a mystery. The mystery was solved when auditory scientists began to study the detailed diffraction of sound waves by the human head, upper torso, and, especially, the external ears (pinna). These anatomical features put their own distinctive stamp on the frequency content of broadband sounds like noise bursts as a function of the relative location of the source. The different diffraction of different frequencies is a directionally dependent filter, and we listeners have got used to the features of that filtering by our own anatomy. When a sound does not exhibit the filtering of our own anatomy, our brains reject the concept that this sound is created by a real source somewhere out there. The appearance of the sound inside the head is apparently a default condition.

Perceived distance: Externalized sounds may be heard as close to us or far away. Understanding how we as listeners perceive the distance of a source is not easy. Of course, when there is a mosquito in your ear, there is a big ILD, but whenever the source is more than a few meters away from the head there are no useable interaural differences between distant sources and nearby sources. A first step in understanding distance perception is to realize that controlled experiments show that we are actually not very good at it. Sound intensity, perceived as loudness, is a cue because of the inverse square law, but when the intensity is randomized, we can't distinguish 2 m from 10 m or more. When sounds are heard in a room, the ratio of direct sound to reverberated sound is a useable distance cue. The ratio is smaller for distant sources, and this is particularly useful in comparing the distance of two different sources in the same room. In addition, very distant sounds are distinctive because the air itself particularly absorbs high frequencies. This filtering effect becomes noticeable when the path length is long.

Compactness: A compact image is perceived to be localized (or lateralized) at a point. The opposite of a compact image is a diffuse image. A diffuse image seems to be spread out in space; it may even surround us. The compact-diffuse distinction is particularly easy to study using headphones. In the early days of sound localization research, experimenters produced tones with ILDs that favored the right ear and ITDs that favored the left ear or vice versa. They wanted to see which interaural difference would win—lateralization to the right or the left? They found that conflicting interaural differences led to an image that was not compact. Sometimes it split in two, other times it became a fuzzy ball inside the head.

Compact images tend to occur when the signals in the two ears resemble one another. There may be an ILD or there may be an ITD, but if the signals are recognizably the same sort of thing on oscilloscope tracings, then a listener will tend to hear a compact image. Monophonic sound reproduction from a single loudspeaker leads to a compact image. Even in a room, a listener can accurately localize the loudspeaker because of the precedence effect. That is not a welcome fact when listening to recorded music. Stereophonic reproduction (two slightly different signals from two loudspeakers) leads to a more diffuse—more surrounding— experience because it helps to make the waves in the two ears to be more different. Similarly, it has been discovered that successful concert halls emphasize reflections from the sides which lead to significant differences in the signals received by the two ears, promoting the surround effect.

Exercises

Exercise 1, The distance effect on ILD
The distance between your ears is about 18 cm. Therefore, if a sound is 1 m away from your right ear, it can be no more than 1.18 m away from your left. (a) Use the inverse square law for sound intensities to show that the level in your right ear can be no more than 1.4 dB greater than the level in your left ear due to the distance effect alone. This calculation ignores the head-shadow effect. (b) Suppose that the source is 0.5 m away from your head. Show that the level difference is larger, namely 2.7 dB. This result generalizes. The closer the source, the greater the ILD caused by the distance effect.

Exercise 2, How big is your head?
(a) Why is it not possible to use the interaural level difference (ILD) to localize sounds with frequencies below 100 Hz? (b) If you wanted to make use of ILD at low frequency would you like your head to be larger or smaller than it is? Why?

Exercise 3, ILDs at 2,000 Hz
Use Fig. 14.2 to estimate the smallest detectable angular change for a 2,000-Hz source. Assume that the source azimuth is between $10°$ and $45°$, and assume that a listener can detect an ILD change as small as 0.5 dB.

Exercise 4, Interaural time difference
(a) Show that Eq. (14.1) is dimensionally correct. (b) Calculate the ITD for azimuths of $30°$, $45°$, and $60°$. Compare your answers with Fig. 14.3.

Exercise 5, Which plane?

Is localization better in the horizontal plane or the vertical plane? What would one mean by "better?"

Exercise 6, The front/back problem

How is it possible to tell whether a sound comes from in front of you or behind you? Such locations would seem to be symmetrical with respect to your ears and head.

Exercise 7, Older listeners

Older listeners have much more difficulty making front–back distinctions compared to left–right distinctions. Can you explain why?

Exercise 8, Precedence effect

What is the precedence effect? Why is it important for localizing sound in rooms?

Exercise 9, Personal experiences

Recall experiences you have had when you were confused about the true location of a source of sound. Sound localization is such a tricky business that everyone has had such experiences.

♠

Chapter 15
Sound Environments

The study of sound in the environment begins by thinking about the different ways that sound can travel from a source to a receiver. There are five important processes as shown in Fig. 15.1:

- *Direct* sound travels in a straight line path through the air from source to receiver. This is always the path of shortest length.
- *Reflected* sound hits a wall or other surface and bounces off. Reflected sound stays in the room with the source. Some of the reflected sound arrives at the receiver in the same room.
- *Transmitted* sound goes directly through a wall to a receiver in another room. Possibly that occurs because there are holes in the wall. Possibly it occurs because the wave from the source vibrates the wall and the wall reradiates the sound on the other side.
- *Propagated and transmitted* sound is like transmitted sound, but it may follow a long path. It may be propagated through air ducts or as vibrations along the surfaces of pipes or partitions. A concrete floor can propagate the noise of a machine long distances within a building.
- *Absorbed* sound is wasted sound. Absorption is different from the other processes because after absorption there is no more sound. Absorbed sound does not get to a receiver, instead it is turned into heat by friction. Actually, the ultimate fate of all sound is to be absorbed somehow, somewhere. That fact is part of a general truth: all energy that is used in our world (mechanical, electrical, chemical, or acoustical) ultimately ends up as heat.

15.1 Reflections from a Surface

The reflection of sound plays an important role in almost all environments. As noted in Chap. 6, reflection from a surface can be specular or diffuse:

W.M. Hartmann, *Principles of Musical Acoustics*, Undergraduate Lecture Notes in Physics, DOI 10.1007/978-1-4614-6786-1_15,

Fig. 15.1 In this cartoon, a source of sound (S) radiates sound that propagates directly to a receiver (or listener, L) by a direct path (D), is reflected from a wall (R), is transmitted through the wall (T), is propagated along the thin panel and transmitted from the end of that panel (PT), or is absorbed on the wall (A)

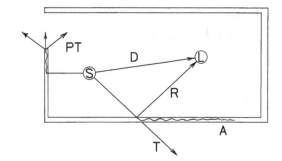

(a) Specular reflection is like a billiard ball bouncing off a cushion on a billiard table. The angle of incidence is the same as the angle of reflection.
(b) Diffuse reflection is like a high-speed tomato hitting a wall. The sound wave bounces off at many different random angles. Specular reflection occurs if the surface is smooth, with bumps that are small compared to the wavelength of the sound. Diffuse reflection occurs if the bumps on the surface are comparable to the wavelength or larger.

Perspective

If you are used to thinking about the way that sound propagates as a wave, complete with diffraction and interference effects, you may be surprised to find sound waves treated as though they were bouncing balls as they reflect from a surface. Does this mean that the nature of sound has changed somehow? No, sound waves are still waves, and there are precise (but complicated) experimental and theoretical ways to deal with the interaction of these waves with surfaces. But for many practical purposes it is an adequate approximation to imagine a wave as following the path of a projectile like a ball as it reflects (bounces) off a surface. The most important assumption in the approximation is that the surface itself (whether it be smooth or bumpy) is large compared to the wavelength of sound. Because sound wavelengths are often large (4 m for bass notes) it is clear that this assumption requires large surfaces. That is why the bouncing ball (or splattering tomato) approach is used in describing rooms. Rooms are large.

15.2   Transmission Loss

We often have the goal of minimizing sound transmission, for example minimizing the transmission of a conversation from one office to another, or the sound of a television from one apartment to another. Unfortunately, the superb dynamic range of the human hearing process works against efforts to reduce the annoyance of transmitted sound. Consider that if you succeed in eliminating 99 % of the sound intensity, you have only reduced the level by 20 dB (see Exercise 4). A reduction of only 20 dB may not be enough to eliminate the disturbance caused by unwanted sound, especially if the sound includes information such as speech.

Sound transmission can be reduced by barriers, such as walls. The reduction of transmitted sound intensity by a wall is the *attenuation* of the wall. A wall can attenuate sound in two ways. It can absorb sound and it can reflect sound. Sound is absorbed by a wall when wall surfaces vibrate and convert the energy into heat through friction. Sound is reflected when it comes to an interface between two different media. A double wall works well because there are multiple changes of medium: air, to wall, to air, to second wall, to air again. Upon each reflection there is some absorption, and that eliminates sound. Walls also transmit sound, and therefore fail to some degree, because of holes in the wall (like a gap under a door) and because the walls are caused to vibrate as a whole by the incident sound. To avoid vibration of an entire wall the wall material should be heavy. Lots of mass is the best way to avoid transmission by vibration. Unfortunately it is a costly way to build.

As a general rule, high-frequency sounds are effectively reflected or absorbed by walls, but low-frequency sounds tend to get transmitted. Plotting the attenuation of a wall (in dB) as a function of frequency always leads to a rising curve—more attenuation for higher frequency. Walls can be characterized by sound transmission class (STC) which refers to the attenuation at 500 Hz, an arbitrarily chosen frequency. If a wall has a STC rating of 40 dB, you can expect that it will attenuate a 500-Hz sound by 40 dB. It will likely attenuate a 200-Hz sound by less than 40 dB and will attenuate a 2,000-Hz sound by more than 40 dB.

15.3 Room Acoustics

The subject known as "room acoustics" is concerned with what happens inside the room containing the source and not about what happens outside. Therefore, it involves direct sound and reflected sound. Sound that is not reflected back into the room is of no interest. Whether this sound is absorbed on the walls of the room and turned into heat or is transmitted out of the room through the walls, the room acoustician says that it is "absorbed." (Ultimately, of course, the lost sound *is* absorbed, even if it happens to be transmitted elsewhere first.) For instance, an open window is said to be a perfect absorber because all the sound that hits the window goes out and never returns. Thus, from the point of view of the room acoustician, sound that hits a wall is either reflected or absorbed. If 30 % of the sound power is reflected, then 70 % is said to be absorbed.

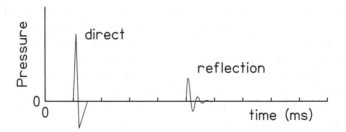

Fig. 15.2 Impulses measured at the position of a listener in a room with a single wall

15.3.1 Early Reflections in a Room

Suppose there is a source, a listener, and a single wall that creates a strong reflection. If the source makes an impulsive sound, like a hand clap at time zero, then there is a brief delay for the direct sound to get to the listener and a longer delay for the reflection (Fig. 15.2).

In a real room with real walls there are multiple early reflections because of single and double reflections from all the walls and other surfaces. Most important for good acoustics is the time gap between the arrival of the direct sound and the arrival of the early reflections. A good listening environment has numerous reflections that arrive with a gap much shorter than 50 ms. Reflections that arrive within the first 20 ms give a sense of intimacy to the acoustical environment. Reflections that come from the side walls give a sense of "surround" or "spaciousness," a desired effect in listening to music. Side wall reflections lead to an incoherence between the sound waves at a listener's left and right ears. The incoherence is responsible for the sense of envelopment—the impression of being "in the sound." Stereophonic sound reproduction has a similar effect in the audio domain. Reflections from the ceiling contribute favorably to the loudness, but they do not lead to the same sense of spaciousness.

Echoes If a strong single reflection arrives more than 50 ms after the direct sound the listener perceives an echo, which is defined as a distinctly heard repetition of the direct sound. Echoes are acoustically bad—really bad. Echoes cause speech to be hard to understand and they cause music to sound "muddy." Even in an environment with lots of other reflections, a particularly intense reflection that is delayed as long as 50 ms can create an echo. Rooms that suffer from an echo must be treated in order to be satisfactory spaces for listening. The location that leads to the offending reflection (usually a back wall) can be covered with absorber. Alternatively, the location can be reshaped to reflect sound over a wide range of angles.

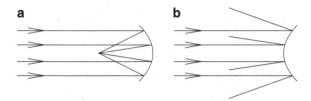

Fig. 15.3 In parts (**a**) and (**b**), sound waves come in from the left and are reflected by a curved surface. In part (**a**) the rays are focused at a point causing the intensity to be large there. In part (**b**) the rays are widely dispersed, helping to create an even distribution of sound intensity

15.3.2 Focused Sound

You know that if sound is reflected from a large, smooth, flat surface, then the sound ray, defining the direction of travel of the wave, bounces off the surface like a ball. Suppose the surface is curved. What then? A concave surface tends to focus sounds as shown in Fig. 15.3a. Concave surfaces are found in cylindrical auditoriums or in rooms with a prominently curved back wall. They occur with domed ceilings as are found in many ceremonial spaces. In extreme cases concave surfaces create whispering galleries, where two people can communicate over amazingly long distances because the vocal waves that are emitted in all directions end up being focused at the location of the listener.

Apart from the fun of whispering galleries, focused sound is normally a bad thing. Focusing of sound can make a delayed reflection strong enough to be heard as an echo. Even if the focused sound is not delayed long enough to create an echo, the focusing concentrates sound power at particular location(s) in the room, while inevitably depriving other locations of power. For good room acoustics, one would prefer that sound power be evenly distributed throughout the room. Even distribution is actually promoted by convex surfaces, which scatter sound waves as shown in Fig. 15.3b. Along with bumpy surfaces, convex surfaces are normally a helpful addition to good sound distribution in a room.

15.3.3 Reverberation

In an ordinary room, sound is reflected from all the surfaces. It is reflected and re-reflected and re-reflected again. After awhile there are so many reflections that it becomes pointless to keep track of individual reflections and where they come from. These reflections come in a dense mass that is called "reverberation." Reverberation is the primary consideration for rooms that are especially designed for listening, and the most important acoustical specification for a room is the *reverberation time*. The reverberation time is the length of time that a tone persists in room after the source has been turned off.

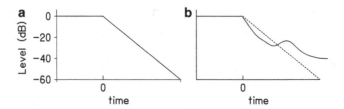

Fig. 15.4 (**a**) To measure the reverberation time we turn off a tone and measure the time that it takes for the sound to fade away by 60 dB as shown by the idealization in (**a**). That time is T_{60}. In the less-than-ideal conditions shown in (**b**) there are two problems. First, the decay is not uniform; second, we can only measure about 40-dB until we reach the noise background. To find T_{60} we use the data we have and estimate the best straight line (shown *dashed*) down 60 dB

The reverberation time is determined by a simple experiment. A tone, with frequency f, is turned on and allowed to fill the room. Then the tone is abruptly turned off. The tone will then decay away and the experimenter measures the time that it takes for the tone to decay by 60 dB. That time is the reverberation time, called T_{60}, at frequency f. The choice of 60 dB is a somewhat arbitrary standard, but it corresponds rather well with our subjective impression of the duration of the decay (Fig. 15.4).

The Sabine Equation It is possible to estimate the reverberation time based on a few simple physical properties of the room. (1) The reverberation time is shorter if there is a lot of absorption in the room, especially on the walls and other surfaces. The greater the absorption, the faster the room loses sound intensity. (2) The reverberation time is longer if the room is larger because sound has to travel further before it can be absorbed on the room surfaces. These two ideas about reverberation time form the basis of the Sabine equation,

$$T_{60} = 0.16\, V/(A_T). \qquad (15.1)$$

Here, T_{60} is the reverberation time in seconds, V is the volume of the room in cubic meters, and A_T is the total absorbing area in square meters. The number 0.16 has units of seconds per meter. It has to account for the fact that the quantity $V/(A_T)$ on the right-hand side of the equation has dimensions of length (meters) and the left-hand side has dimensions of time (seconds). This numerical value (0.16) would be different for a calculation done in English units of volume and absorbing area or for a different reverberation time such as T_{30}.

The volume of the room (V) is a simple concept. If the room is rectangular the volume is just the length times the width times the height. The total absorbing area (A_T) needs some discussion. The A_T is the sum of all the absorbing surface areas in the room. To find an absorbing surface area, you start with the area itself (A), measured in square meters, and multiply by the absorption coefficient (α) for the surface material. Absorption coefficients depend on the material and they depend on frequency too. That is why the reverberation time depends on frequency.

Your intuition already tells you something about absorption coefficients. Suppose a high-frequency tone (e.g. 2,000 Hz) encounters heavy draperies. You expect that a lot of that sound will be absorbed. Therefore it does not surprise you to learn that the absorption coefficient at 2,000 Hz for heavy drapes is 0.70, meaning that 70 % of the sound power will be absorbed. At the other end of the scale, you expect that glazed tile reflects most of the incident sound and you are not surprised to learn that the absorption coefficient is about 0.01, meaning that only 1 % of the incident sound intensity is absorbed. Absorption coefficients can be found in the table at the end of the chapter.

Calculating Reverberation Time The difficult part of Eq. (15.1) is the total absorbing area, A_T. Mathematically, we say that the total absorbing area is a sum of surface areas times their absorption coefficients plus the sum of the effective absorbing areas (EA) attributed to individual objects like chairs or persons.

$$A_T = \sum_{surfaces} A_{surface}\alpha_{surface} + \sum_{objects} EA_{objects} \qquad (15.2)$$

For example, if a room has four walls, each with an area of $100\,m^2$ and an absorption coefficient of 0.3, and a floor and ceiling, each with an area of $400\,m^2$ and an absorption coefficient of 0.1, then the sum over surfaces is $4 \times 100 \times 0.3 + 2 \times 400 \times 0.1$. That sum is $200\,m^2$. If this room is occupied by 40 adult persons then the absorbing area is increased by 40 times the effective area of a person. From the last entry in the Table of Absorption Coefficients we see that the effective absorbing area of a person (at 500 Hz) is $0.45\,m^2$. Thus, the sum of objects is 40×0.45 or 18, and in the end $A_T = 218\,m^2$.

Reverberation and Intensity in a Large Room There is a useful analogy to sound and its reverberation in rooms, namely a stream of water from a hose trying to fill a leaky bucket. The stream of water represents the source of sound. The bucket represents the room, and the leak is the absorption. If the stream of water flowing into the bucket suddenly stops, it takes a while for all the water to leak out of the bucket. The time to drain the bucket is analogous to the reverberation time.

This analogy can go further. Suppose that all the holes in the bucket are somehow plugged up. Then, a continuing steady stream of water into the bucket will cause the level to rise indefinitely. Similarly, if it were somehow possible to eliminate all the absorption in a room, the sound intensity would grow without limit. Eventually the sound would be intense enough to destroy the walls.

In a physically reasonable case, it is not possible to eliminate all the absorption. The correct analogy to realistic absorption is to imagine that the leaky bucket has holes punched up the sides. The higher the water level the greater the flow of water leaking out. What happens then is that the level of water in the bucket will rise until it reaches a steady state where the rate of water flow from the hose is equal to the rate of water flow out all the leaks. Similarly, the level of sound in a room will grow until the rate of sound absorption is equal to the power of the source. As a result, rooms with little absorption are "loud" rooms. The intensity of a steady-state sound in a room from a source with power P is given by

$$I = 4\,P/A_T. \qquad\qquad (15.3)$$

You will notice that this formula for intensity does not include any information about the distance to the source. The formula calculates the intensity in the *reverberant field*, measured at places that are far enough away from the source that the intensity is dominated by reverberated sound and not by direct sound.

Because excessive reverberation makes speech hard to understand, you might imagine that the ideal classroom or auditorium intended only for speech would have no reverberation at all, i.e., the reverberation time would be zero. By aggressive use of absorbers (carpet on the floor, acoustical tile on the ceiling, and sound-absorbing panels on the walls) it is possible to reduce the reverberant sound to a very small level. However, too much absorption leads to the problem of inadequate intensity. For a room seating more than 100 persons, a talker needs reflections from the surfaces of the room to be heard clearly. In the end, when designing a room for speech comprehension a designer cannot simply apply absorption mindlessly. There is no substitute for good judgement.

Recommended Reverberation Times

- For classrooms and other environments where it is important to hear speech clearly, T_{60} should be about 0.5 s.
- For baroque concertos or small jazz ensembles where it is important to hear inner voices, T_{60} should be about 1.5 s.
- For romantic music where the goal is emotional drama, T_{60} should be between 1.9 and 2.2 s.
- To produce the big sound of an organ in a church T_{60} should be 2.5 or greater. A vast cathedral may have a reverberation time as long as 9 s.

15.4   Gaining Control of Acoustical Spaces

When a new concert hall or major theater opens in a community, it is a time of great local interest and pride. It is also a time of anxiety. How will the new hall sound? Will speech be comprehensible? Will music be rich and exciting? The developers or the community have raised millions of dollars to build the hall. Will it work as intended?

You might think that the acoustical design of a hall ought to be a sure thing. After all, the Sabine equation has been known for many years and the absorbing characteristics of materials are well known. There should be no problem with getting the reverberation times right at low, mid, and high frequencies. However, there are other details, especially in the early reflections, that are critical to the success of a good hall. Early reflections have to arrive with the correct delays following the direct sound and with the right density. The balance between low-frequency early reflections and high-frequency early reflections needs to be correct to produce a "warm" sound. It is important that adequate early reflections come from the side walls to establish a sense of envelopment.

In recent years, acoustical consultants have begun to model large and expensive halls on computers as part of the design process. The computer models assume a source (e.g., a piano) on stage and a listener's position somewhere in the audience area. The goal of a computer model is to calculate the transmission of sound from source to receiver and actually let a listener, wearing headphones, hear a simulation of the hall as it is designed. To do that calculation the computer needs a lot of data consisting of the exact location and size of each reflecting interior surface. This information can be taken from architectural drawings. The computer also needs data on the absorption coefficients of the materials chosen by the architect. Once it has all the data, the computer goes to work, calculating thousands and thousands of reflections of the synthetic piano sound. The end result allows a person to put on headphones and hear the sound of a piano being played in the hall before ground is broken for construction. It is obviously easier and cheaper to make changes sooner rather than later. How well does this computer simulation process work? At this point, it is probably too early to say. The technique is still being evaluated. What can be said is that the number of halls that have opened with serious acoustical problems has been less in recent years, compared with only a few decades ago.

Absorption coefficients of materials

Material	125 Hz	500 Hz	2,000 Hz
Acoustical plaster	0.15	0.50	0.70
Acoustical tile	0.20	0.65	0.65
Brick wall—unpainted	0.02	0.03	0.05
Heavy carpet on heavy pad	0.10	0.60	0.65
Light carpet without pad	0.08	0.20	0.60
Concrete—painted	0.01	0.01	0.02
Concrete—unpainted	0.01	0.02	0.02
Heavy draperies	0.15	0.55	0.70
Light draperies	0.03	0.15	0.40
Fiberglass blanket (2.5 cm thick)	0.30	0.70	0.80
Fiberglass blanket (7.5 cm thick)	0.60	0.95	0.80
Glazed tile	0.01	0.01	0.02
Paneling (0.15 cm thick)[a]	0.10	0.20	0.06
Paneling (0.30 cm thick)[a]	0.30	0.10	0.08
Plaster	0.04	0.05	0.05
Vinyl floor on concrete	0.02	0.03	0.04
Wood floor	0.06	0.06	0.06

[a]Plywood paneling supported at 1 m intervals and backed with 5 cm air space

Absorbing areas (m^2)

Object	125 Hz	500 Hz	2,000 Hz
Adult person	0.30	0.45	0.55
Heavily upholstered theater seat	0.33	0.33	0.33

Exercises

Exercise 1, Something there is that doesn't like a wall

Standard home wall construction consists of dry-wall panels mounted on two-by-four studs. Explain the mechanisms of sound transmission through such a wall. How can sound transmission be reduced? Does it help to fill the wall with fiberglass insulation?

Exercise 2, Bad rooms

Think about bad acoustical environments you have experienced. What do you think made them bad?

Exercise 3, Confronting brick walls

You are on a committee to design a new auditorium. Your colleague argues that using a brick interior wall would be preferable to using a plaster wall because the brick wall is rough and will lead to diffuse reflection. Diffuse reflection is desirable in an auditorium. What is your opinion of this argument?

Exercise 4, Decibels once again

The text says that by reducing the sound intensity by 99 % you reduce the sound level by 20 dB. Prove that this statement is true.

Exercise 5, All energy ultimately turns into heat

The text says that the ultimate fate of all energy is to be turned into heat. Is that really true. How about nuclear energy? Would that be an exception?

Exercise 6, Practice with the Sabine equation

In the text, Eq. (15.2) was illustrated with a room having a total absorbing area of 218 m^2. Suppose that the volume of the room is 2,000 m^3. Continue the calculation to show that the reverberation time is $RT_{60} = 1.5$ s.

Exercise 7, More practice with the Sabine equation

Find the reverberation time at 500 Hz of a rectangular room with dimensions 6.3 m by 7.7 m by 3.6 m high, where the walls and ceiling are plaster and the floor is vinyl on concrete. Use the Table of Absorption Coefficients at the end of the chapter.

Exercise 8, Everyone's alternative to Carnegie Hall

Estimate the reverberation time of a shower stall. [Hint Estimate the dimensions of a shower stall and assume that all the surfaces have absorption coefficients similar to glazed tile.]

Exercise 9, Sabine fails for dead rooms

Prove that the Sabine equation fails when absorption coefficients are very high. Hint: Consider the reverberation time when absorption coefficients are 1. Then 100 % of the sound ought to be absorbed.

Exercise 10, Unit analysis

The text says that the quantity $V/(A_T)$ has dimensions of meters. Explain why this is so.

Exercise 11, Expanding the room

Consider two rooms, A and B. The surfaces of room B are made of the same materials as the corresponding surfaces of room A. Thus, the rooms are identical except that every dimension (length, width, height) of room B is twice as long as the corresponding dimension in room A. Compare the reverberation times of rooms B and A.

♠

Chapter 16
Audio Transducers

Audio is the marriage of acoustics and electronics. Audio is used in three major ways: as sound recording and reproduction, as sound reinforcement in rooms or public address, and in broadcasting. In each case, a sound wave is turned into an electrical equivalent (i.e., an electrical analog) and processed in some way as an electrical signal. Then it is turned back into a sound wave. The devices that make the transformation, from acoustical to electrical and back to acoustical, are transducers.

To understand transducers you need to know a little about electricity, and that is where this chapter begins. It introduces the topics of electrical charge, voltage, current, and magnetism.

16.1 Basic Definitions

- *Charge:* Everything in the world is made up of atoms. In the center of an atom is a nucleus that contains protons (positively charged). All round the nucleus are clouds of electrons (negatively charged). In a neutral atom the number of electrons is just equal to the number of protons, and the positive and negative charges cancel. The net charge is zero (neutral). For instance, an atom of copper has 29 protons—that is why it is copper. A neutral copper atom then has 29 electrons too. A neutral atom of aluminum has 13 protons and 13 electrons. Figure 16.1 shows an aluminum atom.

Because of the positive and negative charges, atoms are chock full of forces. (A force is a push or a pull.) These forces are ultimately what gives electricity its prodigious powers. The electrons are attracted to the protons because opposite charges experience a force of attraction. That is why most of the electrons do not wander off into space but stay in the vicinity of the positive nucleus. However, in a material like copper or aluminum, it is possible to separate *some* electrons from their atoms. Some electrons are free to move around throughout the material.

W.M. Hartmann, *Principles of Musical Acoustics*, Undergraduate Lecture Notes
in Physics, DOI 10.1007/978-1-4614-6786-1_16,
© Springer Science+Business Media New York 2013

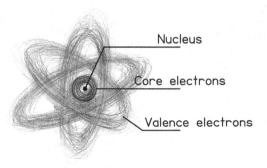

Fig. 16.1 The nucleus of an aluminum atom contains 13 protons, which gives the nucleus a strong positive charge. Around the nucleus are several clouds of electrons. One cloud (the core) is strongly bound to the nucleus. The other cloud (valence electrons) is more spread out. In the aluminum atom there are three valence electrons in this cloud. When atoms of aluminum are put together to make a metallic solid these three electrons are free to move around and to conduct electrical current through the metal

Such materials are called "metals" and they can conduct electrical currents. Electron currents move audio information through wires.

- *Voltage:* Because charges of opposite sign attract one another, a force field is set up when positive and negative charges are caused to be separated. This field of force has an effect on still other charges in the vicinity, and it is known as a voltage. A voltage is measured in volts. For instance, the chemical reaction that separates electrical charges in a flashlight battery leads to a voltage of 1.5 V. These electrical forces, or voltages, can be added together. The basic flashlight contains a stack of two batteries, one on top of the other, leading to a voltage of 3.0 V. The standard 9-V battery is made from six such cells, stacked together in a little rectangular package with snaps on top. You can measure the voltage of a battery by using a voltmeter. It gives a steady reading. Electrical signals, like speech or music in electronic form, consist of voltages that change rapidly with time. You can measure a voltage that changes with time by using an oscilloscope.

- *Current:* Fundamentally, current is the motion of electrical charge. Electrons in a metal are always in motion. All things being equal, they move in random directions—any one direction as much as any other. An electrical current occurs when electrons are caused to move preferentially in one direction, for instance from one end of a copper wire toward the other end. They are caused to move by the force of a voltage. If there is no voltage no current will flow. With a voltage present, either steady from a battery or alternating from an electrical signal, a current flows. For instance, Fig. 16.2 shows voltage from a battery driving a current through a light bulb.

Fig. 16.2 The simple circuit that is used in a flashlight. The two batteries are added together to create a voltage of 3 V. This voltage motivates a current that flows through the light bulb creating light

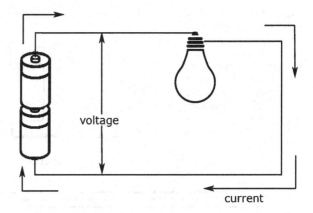

Electrical current in a wire is analogous to the flow of water in a pipe. You can measure the flow of water in a pipe in units of gallons per minute. You can measure the flow of electrons in a wire in units of electrons per second. A more common unit for current is amperes (abbreviation: amps). If you go to the basement of your house and look at the circuit breaker box you will find breakers rated at 15 A. That means that the circuit can safely carry electrical current at a rate of 15 A.

- *Magnetism:* Magnetism is intimately connected with electricity. In fact, every time there is an electrical current there is also a magnetic field. It's a law of nature; it's unavoidable. The reverse is true as well. Every magnetic field owes its existence to some form of electrical current somewhere. (By the way, magnetism is completely unrelated to gravity. People are often confused on this point because our Earth has both a magnetic field and a gravitational field, but these two forces, magnetic and gravitational, have nothing to do with one another, as far as we know.) The relationship between magnetism and electricity leads to the first of three principles of electromagnetism.

16.2   The Current and Magnetism Principle

The current–magnetism principle is:

1. An electrical current in a wire creates a magnetic field around the wire. If the current changes, the magnetic field changes in the same way.

There are two kinds of magnets, temporary and permanent. An electrical current from a battery creates a temporary magnetic field. When the circuit is switched off, the current stops and the field disappears.

Permanent magnets exist because there are certain materials, especially materials that contain iron, that can be made to have a magnetic field because of electrical currents within the atoms themselves. These atomic currents are in a state of perpetual

Fig. 16.3 Current in the coil of the record head makes a magnetic field in the gap. This magnetic field puts permanent magnetism on the tape, which is moving from *left* to *right*. The magnetic field stored at a particular spot on the tape is analogous to the signal in the coil a at particular instant of time

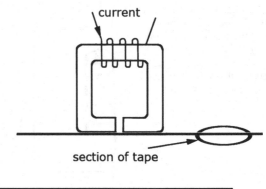

section of tape

Fig. 16.4 A close up of the section of tape from the image above. The *sine wave* represents the input signal, and the *arrows* represent the corresponding magnetic fields that are analogous to the input signal

motion. They never stop. Therefore, permanent magnets can be permanent. You are probably familiar with permanent magnets as compass needles. You can turn a piece of steel (contains iron) into a permanent magnet by putting the steel in a magnetic field, either temporary or permanent. Just being in a strong magnetic field is enough to magnetize a piece of steel—a screwdriver, for example.

16.2.1 Application of the Current–Magnetism Principle

The current–magnetism principle applies to the recording of a sound on magnetic tape. Magnetic tape consists of a mylar backing and a coating (contains iron) that can be magnetized. The coating can be given a permanent magnetic field, just like a screwdriver can be magnetized.

Here's how the recording process works. The acoustical wave is first converted to an electrical signal by a microphone. The electrical signal is amplified and sent to a coil that produces a magnetic field in the record head, because of the current–magnetism principle. As shown in Fig. 16.3, the record head is a solid ring of magnetic material that has a narrow gap where some magnetic field can leak out. As the tape moves from one reel to the other, it makes contact with the record head, right at the gap. The magnetic field in the gap leaves a remnant field on the tape, a permanent record of the signal at a particular instant in time as shown in Fig. 16.4. This is how an audio tape recorder works. This is how a tape camcorder or a VCR works.

16.3 The Analog Concept

The process of recording on magnetic tape, as described above, is an analog process. The signal starts out as a pressure wave in air, and in the recording process it undergoes a number of transformations. It is first converted to a voltage by the microphone, which leads to a current through the wires in the coil of the record head. Then the signal is transformed into a magnetic field in the gap of the record head, and finally it ends up as a remnant magnetism on the tape. The analog concept says that if the process is working correctly then information about the shape of the waveform is preserved in every transformation. If the original pressure wave has a positive bump at some time, then the electrical voltage has an analogous feature at that time, and the magnetism on the tape has an analogous feature at the place on the tape corresponding to the time of the original pressure bump. If the equipment is not working perfectly so that the shape information of the pressure wave is not preserved during the recording transformations then the signal is said to be "distorted." You can hear the difference.

16.4 The Generator Principle

You know what an electrical generator does. It starts with mechanical energy (gasoline engine, steam turbine, waterfall, wind power, etc.) and transforms it into electrical energy. At the basis of that transformation is the generator principle:

2. *If there is relative motion between a wire and a magnetic field, a voltage is induced in the wire.*

Note that the key point is *relative* motion. It doesn't matter whether the wire moves or the magnetic field moves, just so long as one moves with respect to the other. If there is no relative motion there is no voltage.

The generator principle describes the process of playing back a magnetic tape. There is a playback head, very similar in construction to the record head. As the magnetized tape moves past the gap in the playback head, it creates a changing magnetic field in the head. The changing magnetic field passes through a coil wrapped on the head and a voltage is induced in the wire of the coil. This voltage is analogous to the magnetic field on the tape.

The generator principle also describes the operation of microphones of the kind called "dynamic." As shown in Fig. 16.5, the microphone has a diaphragm that is caused to move by the air pressure variations in a sound. A coil of wire is attached to the diaphragm, and so the coil moves with the acoustical wave. The coil is surrounded by a permanent magnet. When the coil moves in the magnetic field, a voltage is created in the coil, and it is this voltage that is the output of the microphone. Positive air pressure causes the diaphragm to move inward, and this creates a positive voltage. Negative air pressure (the small partial vacuum in sound

Fig. 16.5 A dynamic microphone works according to the generator principle. Motion of the diaphragm, caused by the pressure in an acoustical wave, causes a coil to move in a magnetic field. This relative motion causes a voltage to be induced in the coil

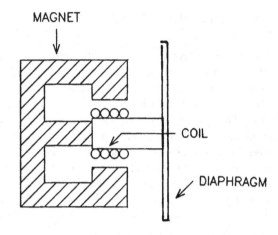

waves) causes the diaphragm to move outward, and this change makes a negative voltage. The voltage can be amplified and used for recording or public address or broadcasting.

The generator principle also is responsible for the operation of a phonograph cartridge, used to play vinyl records. The grooves in the record cause a needle in the cartridge to vibrate. The motion of the needle causes relative motion of a coil and a magnetic field. In this way the cartridge acts as a generator, producing an output voltage.

16.5   The Motor Principle

An electric motor has electrical energy as an input and mechanical energy as an output. It operates according to the motor principle:

3. *If a wire carries a current through an external magnetic field, there is a force on the wire.*

Loudspeakers and headphones operate according to the motor principle. This is true of almost 100% of the loudspeakers of the world, and the great majority of the headphones. We begin with the loudspeaker. A loudspeaker consists of an enclosure—perhaps a rectangular wood box, perhaps the plastic case of a boom box—and sound producing elements called "drivers." We are interested in the drivers.

As shown in Fig. 16.6, a loudspeaker driver has a cone that is attached to a coil of wire. The coil of wire is suspended inside a steel permanent magnet (That is why loudspeakers are heavy.) A power amplifier drives electrical current through the coil and, because of the motor principle, this generates a force on the cone. The force causes the cone to move, and the cone motion moves the air, causing a sound

Fig. 16.6 A loudspeaker driver works according to the motor principle. The interaction of a current (in the coil) with a magnetic field (from the permanent magnet) leads to a force on the coil. This causes the coil to move forward or backward. The coil is connected to the cone which then compresses or expands the air in front of the speaker. You may notice that this figure rather resembles Fig. 16.5 for the microphone. That is not an accident. See Exercise 8 below

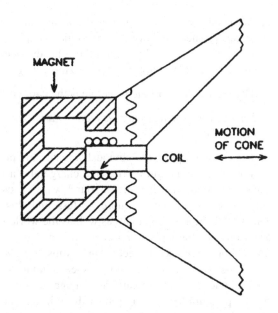

wave. When the current moves in a positive direction, the cone moves outwards, compressing the air in front. When the current is negative, the cone moves inward—into the loudspeaker cabinet—and this creates a partial vacuum. In this way, an alternating electrical signal can create an alternating pressure wave that we hear as sound.

16.6 Electrostatic Devices

Although most audio transducers operate according to the three electromagnetic principles above, some operate according to electrostatic principles. These principles were introduced in the sections above called *Charge* and *Voltage*. There are two principles.

The **fundamental electrostatic** principle:

1. Like charges repel each other, unlike charges attract.

The fundamental electrostatic principle says that two objects with electrical charge will exert a force on one another. Two positive charges repel one another. Two negative charges repel one another. A positive and negative charge attract one another. The electrostatic force has a convenient duality. It can be attractive or repulsive. Therefore, this force can create positive or negative motion (compression or expansion) to make a sound wave in the air.

Electrostatic headphones work according to this principle. Electrostatic headphones are known for their very good high-frequency response because they do not need the heavy magnet and coil that are needed in the usual electromagnetic headphones. However, electrostatic headphones do require a high polarizing voltage—about 100 V. Therefore, these headphones are not so portable. The user has to be close to the supply for polarizing voltage.

The second electrostatic principle is the **capacitor** principle:

2. *Separating electrical charges leads to a voltage. The bigger the separation, the bigger the voltage.*

The capacitor principle obviously creates an opportunity to build a microphone. A moveable diaphragm carries one charge and a fixed frame in the microphone carries the opposite charge. As sound waves cause the diaphragm to move, the charges are separated by more or less and this makes more or less voltage. The world's best microphones as well as the world's cheapest microphones operate according to this capacitor (sometimes called "condenser") principle. The best microphones are used in recording studios; the cheapest microphones are used in telephones and cell phones and for speech input to computer sound cards. The best microphones require a polarizing voltage, normally supplied in recording studios by the preamplifier that amplifies the microphone output voltage. The cheapest microphones are *electret* microphones where the polarizing voltage comes from special materials that have a fixed charge, which creates a fixed electrostatic field much as a permanent magnet creates a fixed magnetic field. (The word "electret" resembles the word "magnet.") Both produce permanent fields, but electrostatic fields and electromagnetic fields are very different. Electromagnetic fields require charges in motion. Electrostatic fields come from charges that are … well … static.

16.7   Electro-Optical Transducers

Electro-optical transducers convert electrical current into light or convert light into an electrical voltage. Electro-optical audio techniques were the first used in putting sound tracks on movie film. The alternative was to separate the sound track from the film images, but the problem of synchronizing sound and pictures in the early days of movie sound made it evident that putting the sound track on the same film with the pictures was the way to go. The sound track was played back by shining a light through the track. The light was then captured by a photosensitive element which created, or modulated, a voltage when struck by the light. The voltage became the audio signal to be played to the audience. The sound track itself was made to pass more or less light by varying its optical density on the film or by varying the width of the track. Analog sound tracks made in this way never worked very well. They were noisy, generated tons of distortion, and had limited dynamic range. The technology of film sound was considerably improved by digitizing the sound tracks.

Electro-optical transducers made from semiconducting materials are now used to play back compact discs and DVDs. They are also used to transmit information along optical fibers. For example, in a compact disc (CD) player, a semiconducting laser (similar to a light-emitting diode) shines a highly focused light beam on the rotating CD. Music is encoded on the surface of the CD in the form of small pits in the surface. When the light beam hits a pit, interference causes the reflected light to be very dim. When the light beam hits a "land" (flat area between the pits) the beam is reflected brightly. The beam, dim or bright, is reflected to a photodiode, which is a semiconducting device that passes current when stimulated by light. Thus, the music, encoded by pits and lands on the CD, causes electrical current to be off and on respectively. The information from the CD is now in electronic form, and that is the most flexible and useful form of all.

These modern optical transducers are different from the other transducers described in this chapter because they do not work on analog principles. Although information is retrieved from a compact disc by modulating a beam of light, there is nothing in the beam of light or the current in the photodiode that is analogous to the music being retrieved. Instead, the compact disc and the modulated light beam convey digital information, a series of ones and zeros. With this technology there is little concern about distortion. Imagine, for instance that a reflected light beam is transmitting a signal that is supposed to be a "one." Suppose there is distortion and the beam transmits the analog equivalent of 0.8. That discrepancy corresponds to a distortion of 20%—intolerably huge for an analog system. Nevertheless, that signal is much closer to 1 than it is to 0, and therefore, the signal will be correctly read as "one." Thus no information is lost by the distortion. More on digital recording appears in Chap. 20.

Exercises

Exercise 1, Basic definitions
Define electrical charge, voltage, and current.

Exercise 2, It's not the size, it's the principle
How is a dynamic microphone like an electrical power generator?

Exercise 3, The origin of magnetism
Is it possible to have magnetism without electrical current in some form?

Exercise 4, Loudspeaker drivers
The motor principle involves a current and a magnetic field. The combination is said to produce a force. This principle applies to loudspeakers. Where does the current come from? Where does the magnetic field come from? What experiences the force?

Exercise 5, Warped images
Putting loudspeakers next to a TV can distortion the TV picture! Why should this happen? [Hint: This is an unintended consequence of the motor principle. Think about the CRT described in the chapter on Instrumentation. Do you see a current

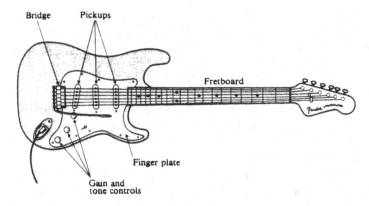

Fig. 16.7 This electric guitar has three pickups—three buttons—for each string

there? Think about loudspeaker drivers. Do you see a magnetic field there? Then there is the force. What experiences the force?

Exercise 6, Where's the motion?

If you look closely at a rock guitar, you will see one or more sets of buttons below each string as shown in Fig 16.7. The buttons are "pickups" that create an electrical current analogous to the motion of the strings. A pickup uses the generator principle to do this. The guitar pickup consists of a coil of wire wrapped around a magnet. How is there relative motion between a this coil of wire and a magnetic field? [Hint: Strings used on an electric guitar are steel, and steel is magnetic.]

Exercise 7, Design a microphone

In a dynamic microphone, the motion of a diaphragm causes relative motion between a coil and a magnetic field. If you had to design a microphone, which would you attach to the diaphragm—the coil or the permanent magnet?

Exercise 8, Motors and generators

It is sometimes said that a generator is simply a motor operated in reverse. (a) Explain how this is true in terms of fundamental ideas of mechanical motion and force and electrical current and voltage. (b) Do you suppose that you could use a loudspeaker driver as a microphone?

Exercise 9, Burned out drivers

The electrical current in the voice coil of a loudspeaker driver, which is necessary to make the cone move, has an unintended consequence. It creates heat. If there is too much current the wire in the voice coil will become so hot that the wire melts. What effect do you expect this to have on the operation of the driver?

♠

Chapter 17
Distortion and Noise

Audio has made it possible to record sounds and reproduce them with reasonable fidelity, to amplify and modify sounds beyond our unassisted human powers, and to broadcast sounds using electromagnetic radiation. Audio has greatly magnified our abilities to control and use sound. It has also led to the twin evils of noise and distortion.

17.1 Noise

When a sound is recorded and reproduced, or when it is otherwise processed or transmitted electronically, it can be contaminated by noise. Noise is something that is added to the signal. You will recognize noise as the "hiss" that you hear on the telephone or radio. You know that the original speech did not contain that hissing sound. The hiss was added because of imperfection in the recording or transmission path of the speech signal. This kind of noise is broadband; it has components at all frequencies. It is caused by randomness in the electronic components because components do not behave in a completely deterministic way. It is a general rule that when signals are processed in their analog form (not digital) then the more complicated the signal path, the more noise one is likely to add. Sometimes the definition of noise is divided into "broadband noise"—meaning the hiss due to randomness, and "hum"—meaning a low-frequency tonal component.

Hum comes from electrical power. Electricity is distributed around the USA as a 60-Hz (60-cycle per second) sine wave, normally with a voltage between 110 and 120 V. It is said to be alternating current (AC) because it alternates from positive to negative. Audio electronic devices are powered by this AC electricity and some of it may leak into the signal path, either due to inadequate design of the equipment or to deterioration of the electronic components. Furthermore, the power lines radiate

W.M. Hartmann, *Principles of Musical Acoustics*, Undergraduate Lecture Notes
in Physics, DOI 10.1007/978-1-4614-6786-1_17,
© Springer Science+Business Media New York 2013

a 60-Hz electromagnetic field, and this can be picked up by cables or by transducers such as dynamic microphones or tape playback heads. Harmonics of 60 Hz, the second at 120 and the third at 180 Hz, also appear in hum.

17.2   Distortion

A signal processing system generates distortion if the waveform coming out of the system does not have exactly the same shape as the waveform going into the system. Imagine a simple acoustical tone that is transformed into an electrical signal by a microphone. We could display that signal on an oscilloscope to determine its exact shape. Now suppose that the signal is recorded onto tape, transmitted through the telephone wires and finally broadcast across the country. We could examine the final signal with an oscilloscope and determine the final waveform shape. If the electronic processing is accurate, the shape should be the same as the original. If it is not, then the signal has been distorted in some way.

17.2.1   Distortion Not

There are several kinds of signal changes that occur in electronic processing that should not be classified as distortion. A simple change in gain, where the signal is multiplied by a constant, is not distortion. This change does not change the shape of the waveform. The effect of the gain change can be compensated by amplification or attenuation.

Inversion, whereby positive is turned into negative and negative is turned into positive, is equivalent to multiplying the signal by -1 and it is not distortion either. The frequency content of an inverted signal is identical to original. Amplification and inversion appear in parts (b) and (c) of Fig. 17.1.

Other kinds of waveform changes are properly called distortion. Distortion may be linear or nonlinear. We deal with these in turn.

17.2.2   Linear Distortion

Linear distortion can be separated into two categories, amplitude distortion and phase distortion.

Amplitude distortion: Ideally, a signal transmission system should have a flat frequency response. That means that it boosts or attenuates all frequencies equally. Linear distortion occurs when the frequency response of the processing system is not

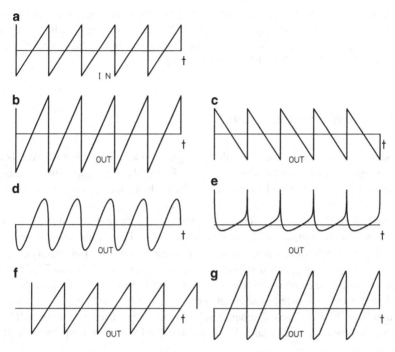

Fig. 17.1 One input and half a dozen outputs from different processing systems: (**a**) The input is a sawtooth wave as a function of time. (**b**) The output of a system with gain—no distortion. (**c**) The output of an inverting system—no distortion. (**d**) The output of a lowpass filter, which reduces the amplitudes of high-frequency components compared to low-frequency components. Consequently the output waveform is smoother than the input—linear distortion. (**e**) The output of a filter that introduces phase distortion. The high-frequency harmonics are just as strong as in the input sawtooth, but their phases are changed, changing the shape of the waveform—linear distortion. (**f**) The output of a time-delay device—no distortion. (**g**) The output of a saturating device that resists extremes, especially negative extreme values tending to cut off the bottom of the waveform—nonlinear distortion

flat—when the system operates like a filter of some sort, giving special emphasis to some frequencies. This effect is known as "amplitude distortion."

Phase distortion: Ideally, a signal transmission system should have no dispersion. That means that if the system delays a signal at all, then it delays all frequencies equally. Such a delay does not distort the shape of a signal. However, if there is dispersion then some frequencies are delayed more than others, then the shape is changed. The change is known as "phase distortion."

Linear distortion is not necessarily a serious matter. Every time you adjust the tone controls on a radio you are distorting (or un-distorting) the signal in a linear way. Linear distortion is part of the normal operation of transducers such as microphones. Theoretically, linear distortion can always be made to go away by filtering the signal. For instance, if the nature of the distortion is to reduce the high frequency content then one can overcome that distortion with a high-boost filter. It may not always be easy to compensate for linear distortion if the processing system is very complicated, but in principle one can always do it. All it takes is a sufficiently specialized filter at the receiving end.

17.2.3 Nonlinear Distortion

Although linear distortion may be rather benign, nonlinear distortion is not. It is the most serious kind of distortion because it can be highly unpleasant and there is no simple way to get rid of it. In general, nonlinear distortion cannot be removed by filtering.

Nonlinear distortion by a device or process results in the addition of extra frequencies to the output of the device or process. That is the most significant difference between nonlinear distortion and linear distortion. For instance, if the input to a device has spectral power at frequencies of 100, 223 and 736 Hz then the output of a linear device will have power at only those three frequencies. The amplitudes may be boosted or attenuated differently, the phases may be scrambled, but there won't be any new frequencies. With nonlinear distortion there will be at least a few new frequencies added, and normally a lot of new frequencies will be added.

The Origins of Distortion Imagine a clock pendulum that is slowly oscillating back and forth, smoothly and unobstructed. Clearly the motion is periodic. It is nearly sinusoidal, but not exactly sinusoidal, and a complete description of the motion would include a few harmonics. Now imagine that someone puts up a barrier so that just before the pendulum reaches its maximum displacement to the right it runs into the barrier—bonk! The pendulum motion continues, it even continues to be periodic, but the pattern of motion is obviously changed by the obstruction—the pattern has been distorted. One might say that the pattern has been "clipped" in the sense that the barrier prevents the pendulum from reaching the full extent of vibration.

Distortion occurs in a loudspeaker driver if you use a finger to prevent the loudspeaker cone from reaching the maximum displacement that the audio signal wants the cone to make. This kind of distortion occurs in a driver, even without interference from the outside, because physical limits in the cone mounting prevent the cone from making very large displacements. If the audio signal sent to a loudspeaker is very large, the signal will try to cause the cone to make extreme displacements and gross nonlinear distortion will occur.

Distortion occurs in audio devices that are purely electronic when the limits of the electronic circuitry are exceeded by a signal that is amplified too much. The usual way to study nonlinear distortion by a process or device is to use one or two sine waves as inputs—thus only one or two frequencies. If there are more, the situation becomes very complicated.

Fig. 17.2 A sine tone suffering symmetrical clipping. Only odd-numbered harmonics (3, 5, 7, . . .) are produced by this distortion

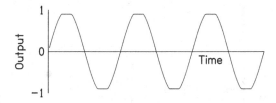

Harmonic Distortion If a sine tone is put into a system that clips off the top there is a change in the waveshape. The tone remains periodic because every cycle of the sine tone has its top clipped off in the same way. Therefore the period is not changed by this clipping, but the signal is no longer a sine tone; it is now a complex tone. The complex tone has a fundamental frequency equal to the frequency of the original sine tone, and now it has harmonics. This is called harmonic distortion. This kind of clipping can be expected to generate harmonics of all orders, both even and odd. The second harmonic or the third will normally be the strongest.

If the nonlinear system clips off the top and the bottom of the sine tone symmetrically, as in Fig. 17.2, then the output includes only odd-numbered harmonics as distortion products.

Intermodulation Distortion If you put two sine tones (frequencies f_1 and f_2) into a nonlinear system, you expect there to be harmonic distortion products, with frequencies $2f_1$ and $2f_2$, also $3f_1$ and $3f_2$, and so on. However, there is more. There are also summation tones and difference tones generated by the nonlinearity. Summation tones occur at frequencies generically given by the formula $mf_2 + nf_1$, where m and n are integers. Frequencies $f_2 + f_1$ and $2f_2 + 3f_1$ are examples.

Difference tones occur at frequencies generically given by the formula $mf_2 - nf_1$, where m and n are integers. The most important difference tones are usually the following three: $f_2 - f_1$, $2f_1 - f_2$, and $3f_1 - 2f_2$. In using these formulas, we consider f_2 to be the larger frequency and f_1 to be the smaller. Normally, then, the formulas will lead to positive frequencies for the distortion tones. However, if one of the numbers you calculate turns out to be negative, you can find a legitimate distortion frequency simply by reversing the sign and making it positive.

Of all these distortion products, the difference tones are the worst. They are easy to hear and are the among the most objectionable features of poor audio. As an example, if two sine tones, with frequencies of 2,000 and 2,300 Hz are put into a nonlinear device, the difference tones $f_2 - f_1$, $2f_1 - f_2$, and $3f_1 - 2f_2$ become, 300 Hz, 1,700 Hz, and 1,400 Hz, as shown in Fig. 17.3.

Distortion in the Cochlea The neural transduction that takes place in the cochlea is a highly nonlinear operation. Therefore you expect that the cochlea generates distortion products. For instance, it should generate harmonic distortion. However, the curious nature of wave propagation in the cochlea saves us listeners from that distortion. You will recall that in the cochlea, sounds travel from the high-frequency end (base) toward the low-frequency end (apex). Suppose that there is a 200 Hz tone that is being transduced by the cochlea. Because 200 Hz is a rather

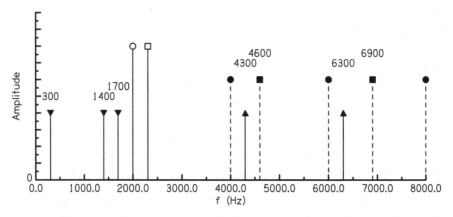

Fig. 17.3 The amplitude spectrum of the output of a nonlinear device when the input to the device is a two-tone signal with components $f_1 = 2,000$ Hz and $f_2 = 2,300$ Hz. The output includes the original frequencies, $f_1 = 2,000$ Hz and $f_2 = 2,300$ Hz, as shown by *open symbols*. The output also includes distortion products: (1) Harmonics of 2,000 Hz, shown by *filled circles*. (2) Harmonics of 2,300 Hz, shown by *filled squares*. (3) Difference tones, shown by *downward pointing triangles*, $f_2 - f_1 = 300$ Hz, $2f_1 - f_2 = 1,700$ Hz, and $3f_1 - 2f_2 = 1,400$ Hz. (4) Summation tones, shown by *upward pointing triangles*, $f_1 + f_2$ at 4,300 Hz, and $2f_1 + f_2$ at 6,300 Hz. There are other distortion components above 8,000 Hz, out of the range of the figure. They include higher harmonics of 2,000 and 2,300 Hz and more summation tones, such as $3f_1 + 2f_2$

low frequency, this transduction takes place near the apex. That's where harmonic distortion products, 400, 600, 800 Hz, ... are created. But in order to be heard these products need to travel to places tuned to these frequencies, and those places are further toward the base where the 200-Hz component is already present and growing. Those places are busy passing the 200-Hz component and respond mostly to it. Therefore, we are not troubled by harmonic distortion.

On the other hand, difference tones have lower frequencies than the input frequencies and they successfully travel to places further along the basilar membrane where the original tones do not create much excitation. Difference tones can be heard. Clear examples of audible distortion products are described in Exercises 6 and 7.

If you want to investigate difference tones created by the ear's own nonlinear distortion, you should start with two sine tones that have completely independent audio paths, from the generators through the loudspeakers. Otherwise, you can't be sure that the distortion you hear is coming from the ear. It might just be coming from the audio equipment. (You can be pretty sure that the air between two loudspeakers and your ears mixes tones from the loudspeakers in a linear way.) An alternative to two independent loudspeakers is a flute duet, where the conditions of Exercises 6 and 7 can easily occur. Audible distortion is actually one of the very serious problems of flute duets.

Distortion in Rock Guitar Distortion is an essential element of the classic rock and roll sound,especially guitar sound. The earliest form of distortion is known as "overdrive," which just means that somewhere along the electronic processing path the signal has been clipped, as in Fig. 17.2. The clipping does not need to be symmetrical, though it often is because early guitar amplifiers clipped in that way and people liked the sound. The clipping can be gentle in which the peaks and valleys are slightly rounded or it can be harsh where peaks and valleys are seriously cut off. Harsh clipping is characteristic of *fuzz boxes*. Guitar distortion is most effective on a strum of all the strings because playing many notes at once, each note with its harmonics, leads to an uncountably large number of intermodulation distortion products and a thick sound. In modern rock guitar sound, distortion in the form of clipping is combined with filtering (often filtering that changes in time) and other time dependent effects such as *phasing* or *flanging*. Fuzz boxes have evolved into "effects" boxes.

17.3 Dynamic Range

The typical electronic signal processor and the typical audio transducer are very linear if the signal is weak. Therefore, it should be easy to avoid the bad effects of nonlinear distortion simply by keeping the signal small. The problem with this approach is that such a small signal may not be large enough compared with the noise added in audio processing. Thus, the signal-to-noise ratio may suffer. The obvious answer to the noise problem is to make the signal larger, but then the system may distort.

A practical example of this dilemma has a singer on stage with a microphone, 50 m away from the microphone amplifier at the back of the hall. If the singer is far from the microphone or sings quietly, the output of the microphone is less than a few millivolts (a few one-thousandths of a volt). The amplifier at the back of the hall needs to multiply that signal by more than 1,000 to get a useful signal of a few volts. In doing so, the amplifier also amplifies the hum and noise picked up by 50 m of microphone cable. That creates a noise problem, characterized by a poor signal to noise ratio. A solution to the noise problem is for the singer to put the microphone close to his mouth and yell. Now the amplifier only needs to amplify by a factor of 100 and the signal to noise ratio improves by 20 dB. Unfortunately, the diaphragm of the microphone cannot readily move to the extreme excursions required by this intense signal source and the microphone causes distortion.

That then, is the nature of the problem. Audio processing must sail the straits between noise on one side and distortion on the other. The width of the straits (to maintain the nautical analogy) is the dynamic range of the system. The dynamic range is the difference, expressed in dB, between the level of an intense signal that causes distortion (say 0.1% distortion) and the level of the noise background. Fortunately, advances in digital electronics have dramatically expanded the dynamic range of audio systems.

Exercises

Exercise 1, Less hum across the pond

European countries use electrical power with a frequency of 50 Hz. Do you expect that "hum" is less audible in European audio gear?

Exercise 2, Too much of a good thing

Why does nonlinear distortion occur if an amplifier is "overdriven."

Exercise 3, Rock 'n' roll

(a) How does nonlinear distortion in the amplifier make chords played on an electric guitar sound "thicker?" (b) If nonlinear distortion makes music sound bad why would the rockers deliberately choose distortion?

Exercise 4, Choose a test signal

You will use your ears to evaluate an audio system for nonlinear distortion. You have a choice of signal sources to put into the system. What do you choose? (a) broadband noise? (b) a complex periodic tone with many harmonics? (c) a sine tone with frequency of 5,000 Hz? (d) a sine tone with frequency of 500 Hz?

Exercise 5, Harmonic distortion

A 1,000-Hz sine tone is put into a nonlinear device causing harmonic distortion. What is the fundamental frequency of those harmonics?

Exercise 6, All the distortion components

You put the sum of two sine tones into a nonlinear device. The sine tone frequencies are 1,000 and 1,200 Hz. What are the frequencies of harmonic distortion products that emerge? What are the frequencies of difference tones that emerge? What are the frequencies of summation tones that emerge? What frequencies are likely to be most audible?

Exercise 7, Worse and worse with up and down

The sine tone frequencies from Exercise 6, 1,000 and 1,200 Hz form the musical interval known as a minor third (See Appendix D if you want to know more about musical intervals.) Suppose the upper frequency is changed from 1,200 to 1,250 Hz. The frequencies of 1,000 and 1,250 Hz make the interval of a Major third. What happens to the difference tone given by $2f_1 - f_2$ when this change occurs?

Exercise 8, Dynamic range

An audio system produces 0.1% harmonic distortion (mostly second and third harmonics) when the level is 90 dB. The noise background is 10 dB. What is the dynamic range of this system?

Exercise 9, Audio amplifier makes toast

This chapter says that electrical power from the outlet in the wall is a 60-Hz sine wave. How is that different from a 60-Hz signal created by a computer sound card and then passed through a power amplifier? If it's not different, why can't you connect a toaster to the power amplifier of your stereo system? (You'd make the connection to the power amplifier just where you would normally connect the loudspeakers.)

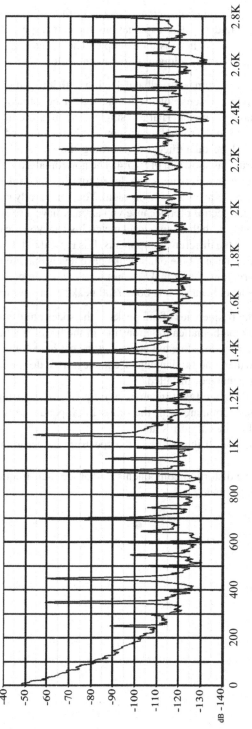

Fig. 17.4 The spectrum of a recorded car horn. (It was a Buick.) The measured sound level was 98 dB, as measured with an A-weighted sound level meter

Exercise 10, Thinking like an acoustician—The car horn

Acousticians try to find physical explanations for human perceptions of sound qualities. The sound of a car horn is an example. The acoustician thinks that the perception can characterized by a loudness and a tone color.

1. Assuming that the spectrum is broadband, having both low and high frequencies, an intuitive approximate sense of the loudness will be gained from a decibel reading on a sound level meter. That is a single, easy measurement to make: Put a sound level meter 3 m in front of a car, blow the horn, and read the meter.

2. The tone color will be quantified by recording the sound electronically and studying its spectrum, as shown in Fig. 17.4. The horizontal axis shows frequency in Hertz. The vertical axis shows the levels of the spectral components in decibels, but the scale is arbitrary. No problem. You don't really need the absolute values of dB on the spectral plot because you already have a sense of scale from the reading you made with the sound level meter. Instead, you are concerned with the relative levels of the different components. It is typical to take the tallest peak as a reference level. Then all the other peaks have levels that are negative. In the case of the car horn, however, it makes sense to take the component at 350 Hz as a standard for the level because it is both a tall peak and a low-frequency peak.

(a) First, consider the frequencies of tall peaks in the spectrum from 0 to 2,800 Hz. Do you see some mathematical relationships in the list of frequencies? Do you see harmonics? What does this tell you about the car horn sound?

(b) If you find more than one fundamental frequency, what does this suggest about how car horns are made or installed?

(c) If you find that there are two fundamental frequencies, determine the ratios of those frequencies. Compare with the ratios in Appendix D to determine the musical interval in the car horn sound.

(d) Can you account for all the peaks in the spectrum by assuming that they are harmonics?

(e) Can you see some distortion components? What are their frequencies and how can you account for them?

♠

Chapter 18
Audio Systems

The purpose of this chapter is to identify some concepts and terms describing audio systems. There is a wide variety of audio systems, ranging from gigantic sound reinforcement systems for stadiums to smart phones and iPods for personal use. Mostly, this chapter will deal with the ideas that are central to practical audio processes. The chapter begins by being amazed about the concept of sound recording.

18.1 Sound Recording

It is quite remarkable to realize that until the middle of the nineteenth century there was no way to preserve sounds. Sounds were produced, had their effect on listeners, and then were lost forever. Any performance of a piece of music had to be a live performance because there was no alternative. The words of orators disappeared as soon as they were spoken, and there was no way to be entirely sure of what was really said because there was no possibility to review. The concept of sound as fleeting—here and then gone—was firmly established, and it must have been a mind-bending experience for people to hear a sound repeated, however badly distorted.

The earliest recorded sounds were a single talker or a single singer. That made sense. Remarkable though the concept of preserving sound might seem, one would certainly expect that a single recording system could preserve and reproduce only a single voice. To produce the sound of a string quartet would logically require four recording systems operating in parallel, and the sound of a symphony would obviously require so many recording systems that they would never fit into an average home. That "logical" idea was rather quickly shot down when recordings were made of singers with piano accompaniment. Preserving the sound of an ensemble in a single waveform stored on some medium dramatically illustrated the fact that the only information that we listeners ever get is a single waveform, a

W.M. Hartmann, *Principles of Musical Acoustics*, Undergraduate Lecture Notes in Physics, DOI 10.1007/978-1-4614-6786-1_18,
© Springer Science+Business Media New York 2013

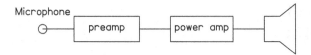

Fig. 18.1 Block diagram of the components of a public address system. In order, they are microphone, preamplifier, power amplifier, and loudspeaker

single pressure function of time (or maybe two—one for each ear), and the business of sorting out the different instruments or voices in the music is all in our heads. Arguably, reinforcing the concept that all the sounds we hear at once are actually summed together to make a single waveform is the most important philosophical contribution of sound recording. Recorded sound continues to amaze.

18.2   Public Address System

The essential elements of audio are embedded in the public address system shown in Fig. 18.1. There is a microphone that converts pressure waves in air into an electrical signal, best described as a voltage varying in time, analogous to the original pressure waves. The microphone may operate according to the generator principle or according to an electrostatic principle, as described in Chap. 16. Then there is an amplifier—a stage of processing that is entirely electronic. Finally there is a loudspeaker that converts the electrical signal back into acoustical form, namely, pressure waves in air. The amplifier in Fig. 18.1 has been shown in two sections, a preamplifier and a power amplifier. One reason for the division is that often these two amplifiers are physically separate units. A more important reason is that the preamplifier and the power amplifier play quite different rolls.

18.3   Preamplifier

The *preamplifier or preamp* is so named because a diagram of the signal path shows that it comes before (hence "pre") the power amplifier. In the public address system, the preamplifier is required to boost the voltage changes that come out of the microphone. The microphone, dynamic or electrostatic, converts sound waves into voltage waves with an amplitude of about $10\,mV$ ($10/1{,}000 = 0.010\,V$). Small voltages like this are not at all nice to work with. They are so small that the inevitable noise and hum that are ubiquitous in the world can badly contaminate them. If one is going to send signals around to different components in an audio system, it is much better to send signals that are at least a volt or two. Signals of this magnitude are called "line level" signals, and power amplifiers and audio signal processors of all kinds typically use signals of this level.

The function of the preamplifier is to provide voltage *gain*, multiplying a weak voltage from a transducer by a factor of 100—maybe even a factor of 1,000—to reach line level. Obviously, there is an advantage to placing the preamplifier physically close to the transducer. A short signal path picks up less hum and noise.

What is true about the small output of microphones is also true about other transducers. Examples are the signals from a phonograph cartridge or a magnetic tape playback head or instrument pickups like the transducers on an electric guitar. In the case of the signal from the phonograph cartridge, the preamp has to apply a filter to the signal, boosting the low frequencies and attenuating the high frequencies, to undo the effects of filtering that are standardly incorporated into the production of vinyl records.

Preamplifiers: Expanded Many years ago, the preamplifiers that amplified the output of phonograph cartridges acquired a second role to play. They were used also for switching among inputs (PHONO, TUNER, AUX, MIC, TAPE-HEAD, etc.) and for tone control (bass, treble, midrange). As audio became more complicated, the multiple channel processing, and functions of surround sound decoding and ambiance synthesis, were also incorporated into the preamplifier.

The back of a modern home entertainment preamplifier bristles with input jacks that accept the outputs from many different sources (PHONO, TUNER, AUX, CD, DVD, VCR, etc.). There may be optical inputs for digital audio. On the front of the preamplifier is a switch that enables the user to select one of the sources. If the user can select more than one source at the same time the preamplifier becomes a *mixer*, which adds several signals.

Of the inputs listed above for the modern preamp, only PHONO is a low-voltage input that requires much voltage gain. The other input circuits are designed for voltage levels of several volts (line level) acquired from other units like a compact disc player (CD), a DVD player, or a tuner. A *tuner* makes it possible to listen to radio stations. It receives AM or FM radio signals, as selected by a tuning device, and puts out the audio signal from the selected radio station.

The Amplification Concept

The ability to amplify, to take a small signal, varying between $+0.01$ and -0.01 V, and multiply its amplitude so that it varies between $+1$ and -1 V (gain of 100), is what makes electronics so useful. Possibly you are wondering how such a device can work. How can a small signal be magnified into a large signal? It seems like getting something for nothing. Fortunately, the processes is less mysterious than it appears.

The essential elements in an amplifier are transistors, which operate like control devices. A transistor is a semiconducting device in a metal or plastic package with three little wires coming out. Two of these wires are connected to a power supply circuit that can easily swing a large voltage, like ± 25 V. The third wire is for a control signal. The essential character of the transistor is that the large voltage swing can be *controlled* by a small input control signal of ± 0.01 V. Changes in the small input voltage control analogous changes in the large output voltage.

The operation of an amplifier is rather like an automobile. You know that pressing down on the accelerator pedal with your foot causes the car to move forward. The more you push on the accelerator the faster the car goes. But the power to move the car does not come from your foot. It comes from the engine, and the pressure of your foot only operates as a control signal. Similarly the output of an amplifier comes from its internal power supply, and the input signal operates as a control.

A transistor amplifier usually includes several successive stages of amplification, each employing several transistors. Alternatively the multiple-transistor circuit may be combined into a single package—an *integrated circuit* amplifier. Before there were transistors for amplification there were vacuum tubes. Tubes operated in the same way in that a large-voltage circuit (output) was controlled by a smaller voltage (input). Vacuum tubes for small signal amplification are now totally obsolete, except for one small niche area; that niche is audio. Some people think that tube amplifiers have a special sound, and they like it.

Modern digital audio techniques set a high standard for signal quality. This standard places great demands on contemporary preamplifiers, which are the first link in the chain after the microphone or other audio transducer. It is a continuing struggle for this part of the technology (an analog part) to keep up, or catch up.

18.4 Power Amplifier

As you learned in Chap. 16, a loudspeaker works according to the motor principle, where the force to move the cone in the driver comes from a current in a magnetic field. It is the role of the *power amplifier* to provide the current. Preamplifiers can't do that. Preamps can produce high enough voltages, but they cannot provide the electrical current or the power needed to drive a loudspeaker.

Power amplifiers for theaters and major home installations are large and heavy because they must supply a great deal of power. Power amplifiers for a stadium occupy an entire room. Inevitably power amplifiers waste some power by converting it into heat which must be eliminated. Therefore, power amplifiers have heat sinks, metal fins that cool the output transistors. Big power amplifiers may require fans for cooling. By contrast the power amplifier that drives headphones or earbuds on your iPod are small and may not be an important source of heat compared to the rest of the electronics. Big or small, a power amplifier is needed any time one wants to convert an electrical signal back into acoustical form.

Normally a power amplifier has a *volume control* that the user can adjust to make the output electrical signal large or small. It may include a *loudspeaker switch* by which the user can select which loudspeakers to drive. The same output may also be used to drive headphones through a *headphone jack*. The headphone jack is connected to the same output as the speakers, through a resistor that limits the amount of current that can be supplied.

The power amplifier used in a two-channel (stereo) system is essentially two identical and independent power amplifiers, one for each channel. Normally, there is a single power supply for these two amplifiers. The stereo power amplifier unit then needs a *stereo balance* control, which operates like a volume control in that it controls the level of the power sent to the speakers. While the volume control raises or lowers the power for both channels, a stereo balance control allows the user to decrease the power sent to the left or right channel in order to achieve a better spatial balance, for instance, getting a vocalist to appear in the center of the sound stage. A multichannel amplifier used in home theater has multiple independent amplifiers, and the correct balance among them depends on the particular loudspeakers in use and the geometry of the environment.

18.5 Mixer

Other analog audio signal devices beyond the basic preamp and power amp can be illustrated by giving our basic public address system a little more capability. For instance, we might like to have live speech from the microphone heard together with music from a source like an iPod. To add two signals together like that, we use a mixer. In its basic form, the mixer simply adds two or more signals. Each of the signals comes in through a channel which has its own gain, g, (volume control)

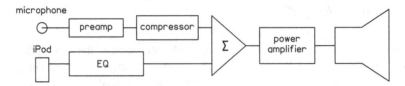

Fig. 18.2 The expanded public address system mixes speech from the microphone and music from the iPod. The microphone signal is compressed and the music is equalized before mixing

so that the user can adjust the mix. You have seen this kind of addition before in Chap. 3. At every point in time the output from the mixer is the algebraic sum of the speech waveform and the music waveform at that particular point in time. That is what the basic mixer does, and that is why the mixer is indicated with a Greek letter sigma (Σ)—the mathematical symbol for addition. It is a linear operation, and it turns out to be very easy to accomplish electronically.

$$x_{out}(t) = g_1 x_{speech}(t) + g_2 x_{music}(t) \qquad (18.1)$$

Large studio mixers are called desks, because they look like huge desks. They are able to add 40 or more channels of sound and to do a lot of processing, equalization for example, on each individual channel. Studio mixers have different outputs, representing slightly different mixes. There will be independent outputs for left and right stereo channels (more for surround sound), submixes to be sent to performers who are playing along with previously recorded sound, and submixes for artificial reverberation devices.

18.6   Tone Controls and Equalizers

Tone controls are filters. Like all filters, they operate selectively on different frequency ranges. A bass control boosts or attenuates low frequencies. A treble control boosts or attenuates high frequencies. To some degree, these controls can compensate for inadequacies in loudspeakers, or the room, or the original sound source, or the personal auditory system of the listener.

In designing our expanded public address system of Fig. 18.2, we have chosen to go beyond the usual bass and treble controls and have selected an *equalizer*. An equalizer can control the level of the sound in individual frequency bands. The variety of possible equalizers is endless. There are five-band equalizers often used in car sound. Ten-band equalizers that boost or attenuate sound in ten different octave bands are standard for home entertainment systems. Professional installations may have 30-band equalizers shown in Fig. 18.3 that boost or attenuate sound in each one-third octave band. Sound reinforcement applications may use *parametric equalizers*, which are individual tunable filters with characteristic frequencies,

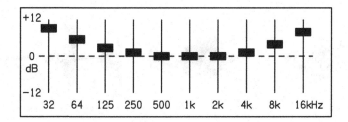

Fig. 18.3 The controls of a graphic equalizer give a visual impression of the filter settings. They are here set in a typical "smiley face" pattern intended to compensate for loudspeaker deficiencies by boosting bands with low and high frequencies

gains, and bandwidths that can be adjusted (tediously) by the audio technician. Individual preferences often play a large role in the way that equalizers are set.

18.7 Dynamic Range Compressor

Dynamic range—the range of levels between noise and distortion—was introduced at the end of the previous chapter. The concept of dynamic range compression begins with the idea that there is some maximum value of the signal that the audio electronics can handle. In radio and TV broadcasting, there is a limit set by law. In analog audio, a high signal value will lead to distortion. In digital audio systems, exceeding the limit creates even worse distortion. This limit has to be avoided.

The negative peaks in the speech signal shown in Fig. 18.4a are right up to the distortion limit shown by the dashed lines, and yet most of the speech is well below that limit. We would like to turn up the gain on the weaker parts of the speech to gain better comprehension, but if we did that, the peaks would become too big. If we could somehow turn up the volume on the entire speech, but limit the heights of the peaks to a safe value we would get speech that is easier to understand. That is the idea of the kind of dynamic range compression known as "peak limiting."

An alternative to peak limiting is compression that boosts the level of weak sounds. Figure 18.4a shows the results of compressing the speech. See how much more intense the weaker parts have become. Dynamic range compression is regularly used in broadcasting. It makes speech sound louder and gives it greater punch. Compression is especially used in commercials. That is why the audio in commercials sounds louder than the audio in the regular program material.

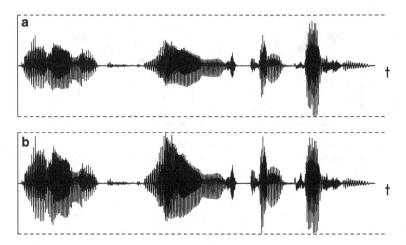

Fig. 18.4 Speech waveform: (**a**) The talker is saying, "Dynamic range compression." (**b**) The same speech is compressed. *Dashed lines* show the distortion limit

18.8 Integrated Amplifiers

As its name implies, an *integrated amplifier* combines a preamplifier and a power amplifier in the same box. The output of the preamplifier becomes the input to the power amplifier. Sometimes the connection between preamplifier and power amplifier is actually made externally with a short jumper. On the back of the chassis is a preamp-out and a power-amp-in. This allows the user to put an external device, a recorder or signal processor, in between the preamp and the power amp if desired.

In an integrated amplifier, the volume controls, stereo balance controls, and tone controls are all part of the power amplifier section. The input switching is part of the preamplifier section. Some tone controls may be put back in the preamp.

18.9 Receiver

The electronic packages that you see most often in the audio stores are receivers. A receiver is an integrated amplifier with a built-in tuner. The internal tuner is one of the inputs that can be selected in the preamplifier stage.

A receiver (or an integrated amplifier) can be used with a recording system, like an iPod or computer sound card by using the output of the preamp and input of the power amp connections on the rear. Such a connection is shown as an "external device" in Fig. 18.5. The output of the preamp goes to the input of the recorder.

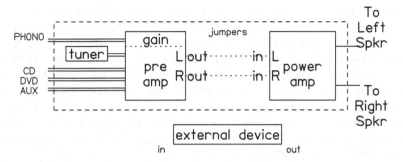

Fig. 18.5 The modules shown within the *dashed box* are the elements of a receiver

Therefore, one records the program that is selected at the preamp. The output of the recorder goes to the input to the power amp. Therefore, one hears the signal that comes off the recorder as processed by the power amp with its volume, balance, and tone controls.

18.10   More Integration

Powered loudspeakers: Increasingly, power amplifiers are being integrated into loudspeakers to create what is known as a *powered loudspeaker* or *active loud-speaker.* Powered speakers, including subwoofers, are discussed in Chap. 19 on loudspeakers. Digital loudspeakers are the same as powered loudspeakers except that they also have a built-in digital to analog converter. Therefore, one can send a digital signal to the loudspeaker directly from a computer or other digital device using a USB connection.

Bullhorns: It is possible to integrate the microphone with the power amplifier and loudspeaker, together with a battery for power, to create a portable public address system, commonly called a *bullhorn* or electronic megaphone. A bullhorn is used to incite, or control, or otherwise influence crowds of people.

Digital microphones: The digital microphone is the converse of the digital loud-speaker. The digital microphone includes its own preamplifier to boost the signal voltage, and it also includes an analog to digital converter to produce a digitized signal right at the output of the microphone. Typically, the digitized signal is sent for further digital processing by means of a USB connection. The main advantage of a digital microphone is that it eliminates the audio cable that runs from the microphone to the preamplifier, replacing it with a digital cable. Because the preamplifier has a lot of gain, noise picked up by the microphone cable can be a problem. In addition, the electronic processing within the microphone itself can eliminate the problem of overdriving the system into distortion with loud sounds.

Wireless microphones: A wireless microphone includes a radio transmitter (often an FM transmitter) to send the signal to a radio receiver that is not very far away. The receiver then sends the microphone signals on for further stages of audio amplification or processing. The transmitter may be a small box that a performer can wear or put in a pocket. Alternatively, the transmitter may be built in to the handle of the microphone. The only advantage of the wireless microphone (but it's a big one) is that the performer is free to take the microphone anywhere on stage or down into the audience without dragging a cable. Because the wireless microphone has potential problems with reliability, radio interference, and audio quality, no one would ever use one except for live performance.

18.11   Multichannel Audio

Stereophonic sound has been around a long time. It was introduced in 1939 in the Disney movie *Fantasia*. Stereo sound became commercially viable in the 1960s with the long-playing vinyl record, and later with stereo FM broadcasting. Stereo is an idea that works in that most listeners find that stereo reproduction greatly enhances the enjoyment of recorded music in comparison with monophonic reproduction. The signals in the two channels of a stereo recording are nearly the same, but they have slight amplitude and timing differences that lead to important directional information. Sometimes, (depending on the room and the listener's position in the room) the directional information reliably localizes different musical sources in the sound field. More important, however, the stereo reproduction leads to an incoherence in the signals received by the two ears which leads to increased transparency in the music (can hear inner voices better) and to an increased sense of envelopment—the impression that the music is all around you.

Multiple channel reproduction (such as a four-, five-, and seven-channel home theater system) attempts to simulate the environment more accurately and provide a more dramatic envelopment by supplementing the usual two front stereo loudspeakers with a center speaker in between them plus at least two speakers in back or on the sides. Again, all the signals in the multiple channels are mostly the same. Larger delays for the signals that go to the speakers in back simulate reflections from a back wall. Changing the delay between front and back speakers can create the illusion that a sound is moving right by you.

Exercises

Exercise 1, Voltage amplification
   Describe briefly how a preamplifier magnifies voltage.

Exercise 2, Preamplifier gain

The chapter says that a preamplifier boosts the voltage amplitude by a factor of 100. (a) Show that this boost corresponds to a level change of 40 dB. (b) What is the level change for a factor of 1,000?

Exercise 3, Designing an equalizer

Graphic octave-band equalizers generally have ten bands. Figure 18.3 is an example. You decide to capture a big share of the equalizer market by producing a graphic equalizer with 11 bands. If the lowest band is centered on 20 Hz, what are the center frequencies of the other ten bands?

Exercise 4, Valves?

About vacuum tubes—the British call them "valves." Why would anyone call them valves?

Exercise 5, What's in it?

Consider the following list of audio elements: CD drive, tuner, microphone, preamplifier, tone controls or equalizer, power amplifier, loudspeakers. Which of these elements do you find in (a) the car sound module that goes in the dash of your automobile? (b) A boom box? (c) A desktop computer? (d) A laptop computer? (e) A camcorder or digital camera?

Exercise 6, Big mixer

The section on mixers suggests a mixer with 40 inputs. Would anyone really ever want to add 40 signals together?

♠

Chapter 19
Loudspeakers

The advances in digital electronics of the last two decades have resulted in enormous improvement in the recording, reproduction, and distribution of sound. The compact disc and DVD have greatly reduced the noise and distortion that were always present when vinyl records and analog magnetic tape were the main distribution media. However, audio is not yet perfect—far from it. The weak links in the audio chain now are the transducers that convert sound from pressure waves in the air to electrical signals or vice versa. Among the transducers, the most problematical are the loudspeakers.

19.1 Loudspeakers: What We Want

Important characteristics that we would like to have in a loudspeaker are these:

1. A linear response

The acoustical pressure wave generated by the speaker should be proportional to the electrical current that drives the speaker. A graph showing pressure vs current should be a straight line. That is the meaning of the word "linear." If the graph is not linear then the speaker leads to distortion.

Distortion causes difference tones to appear due to intermodulation distortion between any two frequency components in the sound reproduced by the speaker. It makes music sound muddy. Loudspeaker distortion tends to be symmetrical, as shown in the Fig. 19.1b. Therefore, distortion products tend to be odd-order.

2. A flat frequency response

The word "flat" means that the graph showing the acoustical output power as a function of the frequency of a tone ought to be a horizontal (flat) straight line. Here we imagine an experiment where we drive the speaker with different frequencies but always the same amplitude. We expect the output power to be constant (flat). In contrast to this ideal, the response for a typical small speaker is shown in Fig. 19.2. It is by no means a straight horizontal line. It has peaks and valleys that cause the speaker to be louder or softer at these frequencies.

W.M. Hartmann, *Principles of Musical Acoustics*, Undergraduate Lecture Notes
in Physics, DOI 10.1007/978-1-4614-6786-1_19,
© Springer Science+Business Media New York 2013

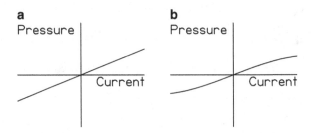

Fig. 19.1 (a) A linear response loudspeaker. (b) A response that is nonlinear in a way that is typical of loudspeakers. It is a *straight line* when the signal is small (close to zero where the axes intersect), but it becomes a *curved line* when the signal is more intense

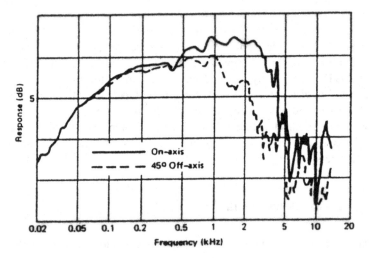

Fig. 19.2 Frequency response of a modern small speaker measured in the forward direction (on-axis) and at 45°. Each vertical division is 5 dB

The frequency response depends on the position of the listener, described by an angle with respect to the loudspeaker's straightforward direction. The reason that the frequency response depends on direction can be found in the ideas of wave diffraction and beaming that were a part of our study of physical acoustics. To review: the diffusion of a sound depends on the wavelength of the sound (λ) compared to the diameter of the speaker (d). The critical comparison is the ratio, λ/d. If λ/d is large (long wavelength–low frequency) then a sound wave spreads out evenly everywhere in front of the speaker. If λ/d is small (short wavelength– high frequency) the wave is *beamed* in the forward direction and does not adequately cover the space to the right or left. Therefore, the problem with direction-dependent frequency response is mainly a high-frequency problem.

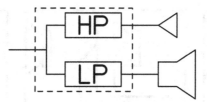

Fig. 19.3 In this two-way loudspeaker, the output from the audio power amplifier comes in from the left and is divided by the low-pass and high-pass filters in the crossover network. The low-passed part goes to the woofer. The high-passed part goes to the tweeter

There is an obvious solution to this problem of beaming of high frequencies, and that is to make the diameter of the speaker very small. Then λ/d is always large and there is an even distribution of sound to all angles. If a speaker were the size of a fingernail there would be no diffusion problem. However, such a tiny loudspeaker would suffer from another problem—it would be very inefficient.

3. Efficiency

Ultimately, the purpose of a loudspeaker is to move air. That is how a loudspeaker transforms power into acoustical form. Efficiency is defined as a ratio of output to input. The efficiency of a loudspeaker is equal to the amount of acoustical power coming out divided by the amount of electrical power going in. If the input power is 2 W and the output power is 0.2 W then the speaker efficiency is 10 %.

Moving a lot of air is especially important at low frequencies where the conversion to acoustical power tends to be inefficient. To move a lot of air requires a large moving surface; it requires a large speaker cone. Then the problem is that a large cone beams high-frequency sound in a forward direction.

19.2   The Two-Way Speaker

Clearly there is a problem. A loudspeaker cone should be small (like a fingernail) to have good diffusion, but it should be large to have acceptable efficiency. A way to solve this problem is to use a large cone for low frequencies—where the wavelength is large, but to use a small cone for high frequencies. Such a system is called a two-way loudspeaker as shown in Fig. 19.3. A two-way speaker receives the entire spectrum of frequencies from the power amplifier, and the division into the low-frequency channel and high-frequency channel is made within the speaker cabinet itself. The low frequencies are sent to a driver called the "woofer." The high frequencies are sent to a driver called the "tweeter." The separation into high- and low-frequency signals is made by a filter called a "crossover." A typical two-way speaker might have a crossover frequency of 1,000 Hz.

A three-way loudspeaker has three drivers, a woofer, a tweeter, and a midrange. The crossover then has three outputs and two crossover frequencies. For instance,

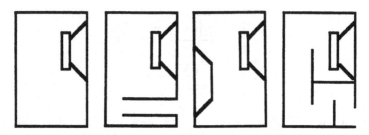

Fig. 19.4 Cross sections of four enclosure types. From *left* to *right*: Sealed box with an acoustic suspension driver. Ported box or bass reflex. Sealed box with added passive radiator or drone cone. Acoustical labyrinth or transmission line

the crossover frequencies might be 500 and 2,000 Hz. Then frequencies below 500 Hz are sent to the woofer. Frequencies above 2,000 Hz are sent to the tweeter, and frequencies between 500 and 2,000 Hz are sent to the midrange driver.

19.3   Enclosures

Imagine a loudspeaker cone coming forward. It is compressing the air in front of it and this makes a pressure pulse. When this pulse arrives at a listener, the listener will hear a drum beat. However, as the front of the cone compresses air, the back of the cone is moving so as to create a rarefaction—a partial vacuum. If air is allowed to move from in front of the cone to the back, some of the pressure pulse generated in front will not reach the listener but will be sucked into the partial vacuum created in back. One might say that the sound has been "short circuited." A way to avoid this problem is to mount the driver in a wall. Then the pressure wave from the front cannot reach the back. The wall is said to serve as a "baffle." If one wants to prevent communication between the front and back of the speaker cone using a portable installation one needs a loudspeaker enclosure to serve as a baffle (Fig. 19.4).

Sealed Box Simple enclosures are sealed boxes. An air-tight box prevents interference between the radiation from the front and back of the driver cone. It is important that the box be rigid, so that vibrations of the box itself do not radiate and interfere with the waves propagated by the drivers. A sealed, rigid box acts like a perfect wall between the front and back of the driver.

Sealed-box enclosures, however, lead to an unintended consequence. They are sealed so tightly that the compression and expansion of air inside the box actually makes a significant contribution to the stiffness of the suspension of woofer cones. To solve this problem, the woofers designed for such enclosures are made with unusually flexible suspensions of their own. It is assumed that the air inside the box will provide additional stiffness. Therefore, loudspeakers made with sealed boxes are often called "acoustic suspension" or "air suspension" speakers.

Bass Reflex (Ported) Enclosure and Alternatives An important design goal for any speaker is to obtain the best low-frequency response (bass response) without using a physically huge loudspeaker. Therefore, an alternative to the sealed box enclosure is the bass reflex or ported enclosure. A port is just a hollow tube that connects the inside of the enclosure to the outside. Therefore, the enclosure is not sealed. The enclosure becomes a resonator intended to reinforce the bass. The length and diameter of the port together with the volume of the enclosure all go to determine the resonant frequency. This frequency is chosen to be just below the lowest frequency that the woofer can easily reproduce by itself. Such a design extends the bass range to lower frequencies.

An alternative to the ported box, also intended to boost the bass response, is the drone cone. The drone cone looks deceptively like a driver, but actually it consists only of the cone with no magnet and no voice coil. Its action is completely passive. The diameter and mass of the drone cone are chosen depending on the volume of the enclosure.

Still another alternative is the acoustical labyrinth, which includes a long path for the radiation from the back of the driver before it meets up with the radiation from the front. The idea of the long path is to delay the back radiation by half a period (path length of half a wavelength) so that it reinforces radiation from the front. Because the goal is particularly to reinforce low frequencies (long wavelengths) such enclosures tend to be large.

19.4   More About Loudspeaker Diffusion

Intuition For a given wavelength, a small-diameter loudspeaker driver tends to diffuse sound over a wide angle, but a large-diameter driver tends to beam sound in the forward direction. Does that seem backwards? Wouldn't your intuition tell you that a small driver ought to beam the sound in a tight angle? This failure of normal intuition on this point is a reminder that the behavior of waves sometimes seems mysterious. The key ingredient that is missing from normal intuition is that the diameter needs to be compared with the wavelength of the sound that is radiated. When the diameter is NOT SMALL compared to the wavelength, the sound is beamed in the forward direction.

Good Frequency Response The fact that loudspeakers don't radiate the same in all directions (they beam the high frequencies) leads to a problem in measuring a frequency response curve like Fig. 19.2. You now realize that if you make the measurement with a microphone directly in front of the loudspeaker (forward direction) you will emphasize the high-frequency response. If you measure with the microphone off to the side you will reduce the high-frequency response. Where should you put the microphone?

The best answer to this question depends on how the loudspeaker will be used. If the loudspeaker will be used in an ordinary living room, you can expect that sound radiated from all angles will be reflected by the walls and ultimately reach

Fig. 19.5 This column loudspeaker is long vertically and narrow horizontally to disperse sound preferentially in the horizontal direction. It is made by stacking five ordinary drivers. Such a stack ends up working just like a long and narrow driver would

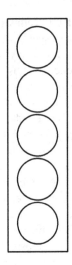

the listener. Sound radiated directly at the listener will be most important, but it will not be *all* important. To estimate the frequency response of the speaker in an environment like this you can put the microphone at an angle of about 45° to the forward direction. This measurement takes some account of the importance of the forward direction but also includes the radiation to the sides that ultimately counts. By contrast, if you are setting up loudspeakers for a concert in the park you can't expect major reflections from walls because there are no walls. Then the forward direction becomes relatively more important.

Intentional Beaming The next time you are in a large church, look at the loudspeakers on the side walls or side columns. You will see that they are long and narrow and oriented with their long dimension in the vertical direction as shown in Fig. 19.5. What's the point of that? Review your new-and-improved intuition about radiation and realize that this loudspeaker will tend to beam the sound (narrow beam) in the vertical direction but to diffuse the sound well in the horizontal direction.

These loudspeakers on the side walls are intended to reproduce speech to make it more intense for the audience and therefore more understandable. By diffusing the sound well to the sides the loudspeakers can cover a wide range of the audience. That's good. But churches, especially large ones, have a problem in that they tend to have high ceilings and vast volumes. The danger in amplifying speech is that sound from the loudspeakers will go up to the large volumes near the ceiling and be reflected back down to the audience. This reflected sound will arrive at the listeners with a considerable delay. That's bad. The original speech plus delayed speech is hard to understand, as described in Chap. 15. The idea behind making the loudspeaker long in the vertical direction is that the speaker will tend to beam the sound sideways toward the audience and not diffuse the sound upward toward the ceiling.

19.5 Powered Speakers

The loudspeakers that have been described so far in this chapter have been passive. They don't need to be plugged into a wall socket. The only source of power they need is the audio power that comes from a power amplifier. There are some advantages to using a *powered* loudspeaker instead. In a powered loudspeaker (active loudspeaker) a power amplifier is included in the loudspeaker enclosure itself. To do this integration, the loudspeaker needs to have its own power supply. One advantage is that the power amplifier can be matched to the loudspeaker drivers. For instance, the amplifier can limit the current to safe levels—protecting the woofer. There can be separate amplifiers for the woofer and tweeter, and the crossover can be improved compared to the passive crossovers used in passive loudspeakers. Even better, special dedicated electronics within the loudspeaker can measure the displacement of the woofer cone and send feedback to the power amplifier to reduce distortion. Powered loudspeakers are commonly used in sound reinforcement applications and as monitors in recording studios.

19.6 Subwoofers

A subwoofer is a loudspeaker specifically intended to reproduce very low-frequency sounds. Therefore, subwoofers are normally physically large. One tends to think of a subwoofer as an "add on." There are several kinds of applications.

Audio System Suppose you have a two-channel (stereo) audio system for listening to music and you want to extend the bass response of the system to lower frequencies. Maybe you want to really experience the 16-Hz organ note that begins *Also Sprach Zarathustra.* Then you can add a subwoofer. But how much you need to reinforce the bass depends on how much bass you already have with your main speakers. Therefore, the subwoofer has a tone control that allows you to adjust its lowpass filter. As you set this filter to higher frequencies you allow the subwoofer to reproduce more of the entire audio range. You are then relying less on your main speakers.

Normally, you would buy only a single subwoofer, intended to reproduce bass sounds from both left and right stereo channels. The advertisements say that it does not matter where you locate the subwoofer because low-frequency sounds are "not directional." That may be true if the subwoofer is set to reproduce only the very lowest frequencies. But many installations require boost at frequencies as high as 100 Hz. If the subwoofer reproduces sounds with frequencies of 100 Hz or above, the directional characteristics *do* matter, and using a single subwoofer would be a compromise.

For your audio installation you can buy a passive subwoofer or a powered subwoofer. The driver in a passive subwoofer has two independent voice coils, one for the left channel and the other for the right. The two voice coils are wound together within the same magnet, and so the subwoofer cone is moved by the sum of the left and right signals. In operation, you send both the left and right channel outputs from your power amplifier or receiver to the subwoofer, which combines them electromagnetically as described above and then sends separate highpass-filtered versions of the left and right signals to the main speakers.

Because a powered subwoofer includes its own power amplifier, you are not relying on your power amplifier or receiver to provide the power to drive the subwoofer. An additional advantage of the powered subwoofer is that electronics within the subwoofer itself give you more precision and flexibility in the lowpass filtering that selects the frequencies to be reproduced by the subwoofer.

Sound for Television Home entertainment centers combine commercial television reception with video tape, DVD recordings, and digital recorders. Control centers, or video receivers, include enough separate power amplifiers to drive arrays of passive loudspeakers like a 5.1 array. Here, the number 5 refers to the number of full-range loudspeakers: left and right speakers in front, left and right speakers in back, and a center speaker in front. That makes five. The 0.1 refers to a subwoofer channel which reproduces only low frequencies. A 7.1 system is like a 5.1 system but adds two speakers on the sides. To set up such a system you need to set the band of frequencies that will be sent to the subwoofer. As for the musical application, the setting depends on the capabilities of your main left and right front speakers. The receiver setup procedure asks you whether the main speakers are large or small to make that decision. In installations with high-quality main speakers the subwoofers mainly reproduce only "low-frequency effects (LFE)" such as explosions or the rumble of engines.

Sound for Computers Loudspeakers attached to computers are powered loudspeakers to reduce the power load on the computer and its sound card. Typically computer loudspeakers are small. Perhaps they are small because computers are used on office desks and there is never enough room on a desk. Perhaps it is because of price pressure in computer sales. Perhaps it seems wrong for a computer loudspeaker to be larger than the computer itself. For whatever reason, computer speakers are normally small, and the sonic quality is a compromise. To try to improve the sound, one may add a subwoofer, normally placed somewhere out of the way. The subwoofer cabinet may also contain the power amplifiers for the main left and right speakers, or for a complete set of both main and rear speakers. In a commercially available computer sound system with a subwoofer, the subwoofer typically handles a great deal of the bass.

Exercises

Exercise 1, Engineering = compromise

Describe how different goals of loudspeaker design lead to compromises. One goal is high efficiency. Another goal is a frequency response that does not depend on direction.

Exercise 2, Building an enclosure

Suppose you are building a bass reflex speaker using a woofer that rolls off (limit of low-frequency reproduction) at 60 Hz. What would you choose for a resonance frequency for the cabinet and port?

Exercise 3, A three-way loudspeaker

Draw a diagram of a three-way loudspeaker showing the three drivers and the crossover—with low-pass, band-pass, and high-pass filters.

Exercise 4, Efficiency comparison

Which is more efficient? Speaker A has an output power of 1.5 W given an input of 20 W. Speaker B has an output power of 1 W given an input of 14 W.

Exercise 5, Understanding the spec sheets

Loudspeaker efficiencies are normally quoted by giving the output sound level in dB SPL (sound pressure level above threshold) measured at a distance of 1 m, for an electrical input power of 1 W. Using this measure find the efficiency (in dB) of speaker B from Exercise 4? (Use the inverse square law.) Remember that the threshold of hearing is 10^{-12} W/m^2.

Exercise 6, Linear for small amplitudes

The caption to Fig. 19.1 says that the nonlinear response in part (b) is actually linear when the signal is small. Compare the slope of that linear part with the slope in part (a). What do you conclude in comparing parts (a) and (b)?

Exercise 7, Frequency response

For the loudspeaker in Fig. 19.2, the frequency response for the forward direction (0°) is about the same as for 45° up to about 500 Hz, but the responses for those two angles become quite different above 500 Hz. Is that expected behavior? Can you imagine an opposite situation where the responses are similar at high frequency but not at low frequency?

Exercise 8, Low-frequency response

For the loudspeaker in Fig. 19.2, compare the response at 50 Hz to the response at 500 Hz. By how many decibels has the response "rolled off" at 50 Hz?

Exercise 9, Multi-channel loudspeaker arrangements

Figure 19.6 shows loudspeaker arrangements of increasing complexity. Count the number of loudspeakers required for each arrangement. Estimate the costs in money and floor space. Where do you draw the line between costs and sound surround?

Fig. 19.6 Loudspeaker
arrangements for home
entertainment systems.
Squares indicate speakers.
The *circle* indicates a listener.
(**a**) Stereophonic. (**b**)
Stereophonic with mixed *left*
and *right* channels sent to a
speaker in the *middle*.
(**c**) Quadraphonic. (**d**) 5.1
system. [The code 0.1
indicates the
low-frequency-effect (lfe)
subwoofer.] (**e**) 7.1 system

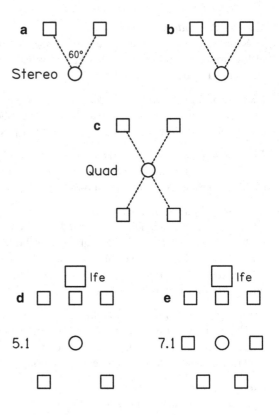

Chapter 20
Digital Audio

In the past several decades, digital electronics, as represented by computers, have made great changes in the way we do things. Audio is no exception. The purpose of this chapter is to describe important principles involved in digital electronics and its application to audio. It especially contrasts digital audio with analog audio.

20.1 Digital vs Analog

Even in a digital world, audio begins and ends with an analog signal. The output of a microphone or other acoustical transducer is an electrical signal that is analogous to the acoustical waveform that produced it. The signal remains in analog form until converted to digital form by an analog to digital converter (ADC). Once in the digital domain the sound can be transmitted, processed, or stored as digital data, i.e., as a series of "ones" and "zeros." To hear the sound again requires that it first be transformed back to analog form by a digital to analog converter (DAC). The analog signal can then be amplified to produce an analogous current that will drive loudspeakers to make an analogous acoustical signal. What is different about digital audio is the digital means of transmission, processing, and storage.

Transmission: The sound that comes over the Internet, or in digital broadcasting, or DMX digital cable is digital transmission. It is transmitted as a series of high and low voltages, representing ones and zeros. In this way it is different from analog transmission, as in AM or FM radio, where voltages follow the original acoustical material.

Processing: Digital processing exploits two splendid advantages of the digital domain, the availability of delay and noise immunity.

Digital methods can delay a signal by any desired amount of time because the signal is readily available in digital memory. There it can be processed in detail, like any other digital file, by a computer or a computer-like device. An obvious

W.M. Hartmann, *Principles of Musical Acoustics*, Undergraduate Lecture Notes
in Physics, DOI 10.1007/978-1-4614-6786-1_20,
© Springer Science+Business Media New York 2013

application is an ambiance synthesizer, where a signal (music or speech) is delayed and added back many times to the original to simulate the reflections in a room. The delayed signal can be delayed by a small amount to simulate early reflections and it can be given many different long delays and added back to simulate reverberation. The delayed signals can be attenuated and can be filtered to represent the absorption of high frequencies on reflecting surfaces. Less obvious, a delay operation is at the heart of digital filtering, which can change the spectral profile of a signal, just like an analog filter, but with more precision and with more flexibility.

An analog signal processor—an analog filter for example—is a *causal* device. Its output at any given time depends on the input signal at that time, or at times in the past. Such a processor cannot make any use of the signal as it will be in the future because it has no knowledge of the future. This seems to us to be a reasonable limitation. We ourselves live with this limitation every day. It is a constraint called "causality." Digital processing devices do not have to live with this constraint. An input signal can be stored in memory, processed using all the information in the memory, and then reproduced—with a time lag that depends on the memory length, of course, but normally only a few milliseconds. Thus, digital devices can be *acausal*, meaning that what happens now can be affected by what will happen in the future. Freedom from causality opens up an entire new world of signal processing opportunities.

Noise immunity is an important advantage of processing in the digital domain, because the analog domain is not immune to noise. As analog processing increases, noise and distortion are added—combining signals in a mixer, filtering them to improve frequency response, filtering again to eliminate some problem, running the signals across the room in cables, etc. Each added stage or process degrades the signal somewhat. This limits the number of operations that can be chained. It particularly limits the number of stages (hence the precision) of filtering that can be done in the analog domain. By contrast, digital processing is immune to noise. Adding another digital process to a signal path, or adding ten more digital processes, does not increase the amount of noise or distortion. That is because digital operations are numerical. Exercise 1 illustrates this point.

Among the most important processing operations is the simple operation of copying. With analog recordings there is some signal degradation with each successive recording. With digital recording and storage there is no degradation from one generation to the next. (See Exercise 13.)

Storage: Compact discs, DVDs, digital audio tape (DAT), diskette, and hard drives are examples of digital storage media. So are computer RAM and flash drives. In their physical principles, some of these storage devices are not very different from the devices that have traditionally stored sound in analog form. The compact disc, or DVD, uses optical storage, resembling the optical storage of sound tracks on film. The DAT, the hard drive, and the diskette are magnetic media, storing digital data as small regions of magnetism on the medium, just like a tape recorder stores analog data. What is different about the digital storage is not the physical principle of storage; it is the matter of *what* is stored.

Digital data are stored as numbers. The numbers are represented in binary form. Each binary digit (bit) is either a "1" or a "0." The compact disc is a format that uses a 16-bit word. Therefore, an example of a meaningful sample of a waveform stored on a compact disc might be

0110 0111 1101 1001.

If this number is recorded on a magnetic disk, then successive places on the disk are magnetized so:

where the arrow points in from the north pole to the south pole of the magnet.

This number can be interpreted from right to left: There is 1 in the "1"s column, zero in the "2"s column, zero in the "4"s column, 1 in the "8"s column, and so on. The four bits on the right add up to $1 \times 1 + 0 \times 2 + 0 \times 4 + 1 \times 8 = 9$. With 16 bits it is possible to encode $2^{16} = 65536$ different numbers. (See Exercise 2.) A problem immediately arises because all the numbers are positive and we wish to store waveforms that are both positive and negative. This problem is solved by a rule, applied in audio, that says that if the first bit on the left is a 1 then the number shall be interpreted as negative. Hence the number above is positive because the first bit is "0". (Digital storage conventions are full of arbitrary rules like that. Digital methods work only because everybody agrees on the rules.)

It now becomes possible to see why digital storage is immune to noise. When a bit is stored on a hard drive the magnetism will point up or down. Of course, there will be random variations in magnetism on the disk. At some places the magnetism will be slightly larger than other places because of inevitable noise on the magnetic medium. So long as this variation in magnetism is small compared to the difference between upward and downward magnetism, the noise will not cause the direction of the magnetism to be misinterpreted. A "1" will remain a "1," and a "0" will remain a "0." The digital information is not corrupted at all by this noise. The information is preserved perfectly as shown by Fig. 20.1. This immunity can be contrasted with analog storage where any added noise is reproduced along with the signal.

20.2 Digital Noise

There is, however, a form of noise that occurs particularly in digital storage, namely quantization noise. A 16-bit word allows only 65536 different possible values of the waveform at an instant. It cannot store values of the waveform that happen to lie between allowed values and therefore it makes an error. Since allowed values are 1 unit apart, the largest possible error is 1 unit. This idea leads to a calculation of signal to noise ratio (SNR). The calculation goes as follows: First, the quantization

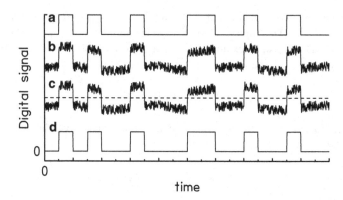

Fig. 20.1 (a) A 20-bit digital word, [0101 0010 0011 0010 0100], is put out to a storage device.
(b) The storage device records the word as a signal, adding hum and noise. (c) The recorded
signal is interpreted with a threshold set at the *dashed line*. Whenever the signal is greater than the
threshold, the signal is interpreted as a 1; else it's a zero. (d) The interpreted signal is recovered
from the storage device. It represents the word perfectly

noise is 1 unit. Second, the largest possible signal is 65536 units. Therefore, the
SNR is 65536/1, and in decibels this becomes $20 \log(65536) = 96$ dB. This is an
excellent ratio of signal to noise. The best vinyl records achieve a ratio of only 60
or 70 dB. In general the SNR in dB can be written as

$$SNR = 20 \log(2^N) = 20\,N\,\log(2) = 20 \cdot 0.301 N \approx 6N. \qquad (20.1)$$

This formula, as developed from left to right, says that for every extra bit in the
digital word one gains 6 dB of SNR. To understand the factor of 20 that multiplies
the logarithm, please refer to Exercise 8 in Chap. 10 that shows how to deal with
amplitudes on the decibel scale.

As usual, it is hard to illustrate realistic noise with a drawing. The ear is so
much more sensitive than the eye in this matter. To illustrate quantization noise one
can imagine what would happen if the word length were 3 bits instead of 16 bits.
Figure 20.2 shows a sine signal and the sampled sine signal using a 3-bit *rounding*
DAC. The rounding DAC replaces a number by the nearest integer value. The
quantization noise would be even worse with a *truncating* DAC. A truncating DAC
replaces each number by the integer part of that number. For instance, at the time
of the second sample in Fig. 20.2 the signal has a value of 2.7. The rounding DAC
replaces this value by 3. The truncating DAC replaces it by 2.

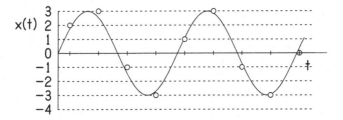

Fig. 20.2 A signal with amplitude of 3 units is shown by a *continuous line*. It is sampled regularly at times shown by *tic marks* on the *horizontal axis*. The sampled version of the signal is shown by *circles*. The analog-to-digital conversion is only 3 bits and so there are only 8 possible different values $-4, -3, -2, -1, 0, +1, +2, +3$. The resulting quantization error is evident in the figure. The *circles* do not agree perfectly with the *solid line*

20.3 Sampling

When sound is recorded on analog magnetic tape or vinyl records it is recorded continuously. A digital recording cannot be continuous. Instead, a digital recording consists of a series of samples of the sound. These samples are taken by the ADC at a precisely regular rate. Figure 20.3 shows such sampling.

Obviously, if the samples are taken too infrequently important details of the waveform will be missed. In order to create a faithful sampled image of the waveform, the samples must be taken frequently. But how frequently do the samples have to be? If the waveform has only low frequencies so that it changes only slowly, the sample rate does not need to be high. But high frequencies will cause the waveform to change rapidly and a high sample rate will be needed to capture those rapid changes. In the end, the correct sample rate to use depends on how high the highest frequencies are. The rule is called the Nyquist rule.

> **The Nyquist Rule:**
> **The sample rate must be at least twice as high as the highest frequency in the signal to be sampled.**

The sample rate for compact disc is 44,100 samples per second (sps). Therefore, the sound to be recorded digitally cannot have a frequency greater than $44,100/2 = 22,050\,\mathrm{Hz}$. That's OK. The audible range extends only to 20,000 Hz. The sample rate for DAT is slightly higher than for CDs. It is 48,000 sps.

20.4 Contemporary Digital Audio

Digital techniques are so flexible technically that they lead in many different directions. We note some of them here:

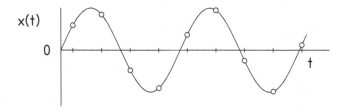

Fig. 20.3 The signal from Fig. 20.2 is sampled regularly at times shown by *tic marks* on the *horizontal axis*. The analog-to-digital conversion is 16 bits and this sampling is so accurate that it is not possible to see the difference between the true signal, shown by the *solid line*, and the samples, shown by the *circles*. Some people think that they can hear the difference though

Improved quality: The compact disc standard—16 bits and 44.1 ksps—leads to high quality sound, but there are many who think that it is not good enough. Already 24-bit recording is standard practice in digital studios. The DVD medium supports both longer word lengths—such as 24 bits—and higher sample rate. The most important reason for the flexibility of the DVD is that its storage capacity is nearly seven times greater than the CD.

Data compression: The compact disc standard requires a great deal of data (see Exercise 3). If there are 700 MB on a compact disc, then a 32-GB computer flash drive can comprehend the contents of only 45 compact discs. The problems of great amounts of digital data become really serious when transferring audio over the Internet. The slow speed of many links lead to long download times.

Reducing the amount of data that is required to transmit and store audio, while minimizing the degradation in audio quality is the object of data compression techniques. These techniques work by taking time slices of the original material, making a spectral analysis of each slice, and processing slices with a data reduction algorithm based on what listeners are likely to hear. Then the slices are joined back together to make a digital audio stream with fewer bits per second. Data reductions of one bit *out* for every ten bits *in* are possible with techniques such as MP3 encoding. The operation of compressive encoding schemes demands that the playback process be compatible with the encoding process. Tight standards must be in force.

Convergence: In the days of vinyl microgroove records there were just a few formats. Long-playing records were made at 33 RPM (revolutions per minute) and the disks were 10 in. or 12 in. in diameter. Singles were recorded at 45 RPM, and the disks were 7 in. in diameter. There was a clear reason to restrict the number of formats because introducing a new format required new hardware on the part of the user. Users resisted frequent changes of hardware.

The matter of formats is entirely different with digital media. A new format may require only a new program for interpretation, and new programs are easily and cheaply distributed—for instance by the Internet. There is no compelling reason to limit audio to two channels (stereophonic), or four, or six. Audio can be combined with video and other data. Programs in different formats can be accommodated on the same distribution medium so long as the instructions for decoding are included. The combination of all forms of communication in a comprehensive and flexible data space has been called "convergence." Convergence requires that competing manufacturers, and competing distributors of program material, and governments agree on standards to be followed by the industry as a whole. With all interested parties jockeying for position, it is sometimes a slow-moving process.

Exercises

Exercise 1, The joy of integer numbers
You know that $1 + 1 = 2$. That is an accurate statement. How much less accurate is the statement: $1 + 1 + 1 - 1 + 1 - 0 - 1 = 2$? The answer to this question explains why digital signal processing can be chained indefinitely with no degradation in signal quality.

Exercise 2, Test the formula yourself.
A formula in the text says that with an N-bit word, it is possible to encode 2^N different numbers. This statement clearly works for $N = 2$. The formula gives 2^2 and this is equal to 4. It is easy to write down all possible combinations of two bits:

$$00 \quad 01 \quad 10 \quad 11,$$

and we discover that there are four possibilities as predicted by the formula. Show that the formula works for $N = 3$ and $N = 4$. This exercise is intended to persuade you that the formula is correct for any value of N. It is an example of *inductive* logic.

Exercise 3, Compact discs
The sample rate for compact discs is 44,100 samples per second per channel. Each sample is stored as a 16-bit word. Show that it requires at least 1.4 million bits to store a single second of stereophonic sound. There are 8 bits in a byte. How many bytes does it take to store a single second of sound?

Exercise 4, Nyquist theorem
Explain why the Nyquist theorem says that there must be at least two samples per cycle of a waveform.

Exercise 5, Digital telephones
Signals transmitted over the telephone are normally restricted to a maximum frequency of 5,000 Hz. What is the smallest sample rate that can be used to digitize a telephone conversation?

Exercise 6, No apparent audio source
The first section of this chapter explained that digital audio required an ADC to convert analog signals into digital form before the signals could be processed and stored using digital methods. Can you think of an exception to this statement?

Exercise 7, Communicating with extraterrestrials

We are going to send a capsule into deep space with artifacts from our civilization with the expectation that someone out there will find it and learn about our human culture. It is the space-age equivalent of the message in a bottle tossed into the sea. We want to include an audio recording of some human speech and music. What recording format do you recommend, a digital compact disc or an analog vinyl record?

Exercise 8, Digital ready?

In the early days of digital audio, some loudspeakers and headphones were marketed as "digital ready." What might that mean?

Exercise 9, Quantization noise?

Why is it called "quantization noise?" Why isn't it called "quantization distortion" instead?

Exercise 10, An 8-bit system

Early digital audio systems used 8-bits. Quantization by 8 bits is still used in some digital applications such as digital oscilloscopes, but 8-bits is not good enough for audio. There is too much quantization noise.

(a) How many different possible values can be represented by an 8-bit system?
(b) Calculate the SNR of an 8-bit system. Do you agree that this is not good enough for audio?

Exercise 11, An 8-bit word

A byte is an eight-bit word. Show that the byte below represents the number 109.

0 1 1 0 1 1 0 1

Exercise 12, Have another byte

Write the 1s and 0s in a byte to represent the number 85.

Exercise 13, Sharing commercial music

Music in digital form can be copied, recopied, and copied again many times with no loss in quality. What are the implications for the industry that produces and distributes music?

Exercise 14, Causality

Making a decision now that is influenced by events that will happen in the future violates the principle of causality. (a) Recall an instance in which you wished that you could violate causality in this way.

Violating causality risks a time paradox where making a decision based on the future changes the future in such a way that the decision is no longer based on the future. (b) Does your recalled instance create a time paradox?

(c) The first section of this chapter says that digital signal processing can violate causality. Really? Explain why it does not involve a time paradox.

♠

Chapter 21
Broadcasting

The soprano sings at the Metropolitan Opera at Lincoln Center in New York, and the performance is broadcast by National Public Radio. Back at home you take off your Gucci slippers, relax in your hot tub and listen to the performance. How does it work? One thing you know for sure—the sound is not transmitted from the Met to you as an acoustical wave. Instead, the sound is transmitted as a radio wave, a form of electromagnetic radiation.

It is sometimes said that a radio station is "on the air," or that radio broadcasting uses the "air waves." This is a figure of speech; technically it is nonsense. The air has nothing to do with it. Like all forms of electromagnetic radiation, radio waves travel perfectly well through an absolute vacuum. It's a good thing too. All of the energy that we get from the sun arrives here as electromagnetic radiation coming through the vacuum of space. Electromagnetic radiation travels at the speed of light, and therefore the performance from the Met comes to you at the speed of light. The speed of light is

$$c = 3 \times 10^8 \text{ meters per second,} \tag{21.1}$$

which you will recognize as about one million times faster than the speed of sound.

Practical radio waves have frequencies from about 3,000 Hz to 100 GHz. That is, from 3×10^3 to 10^{11} Hz. You will note that the lowest frequencies are in the audible range. Do not be fooled. These low-frequency radio waves are electromagnetic; they are not acoustical—they cannot be heard. Still, radio waves are indeed waves, and they obey many of the same laws as acoustical waves. For instance, the famous relationship between wavelength and frequency still holds,

$$\lambda = c/f, \tag{21.2}$$

where in this case, c is the speed of light.

W.M. Hartmann, *Principles of Musical Acoustics*, Undergraduate Lecture Notes in Physics, DOI 10.1007/978-1-4614-6786-1_21,

Another important feature of radio waves that we might anticipate from the study of acoustics is the phenomenon of difference tones, generated in a nonlinear system. We will see that presently.

21.1 Continuous Wave

A radio transmitting system consists of a source of electromagnetic power in the form of a sine wave with a precise frequency in the radio range. The source of power is called a transmitter, and it is attached to an antenna that radiates the sine wave. For instance, the radio wave from Station X might have a frequency of 1.000 MHz. We would say that Station X is broadcasting at a frequency of 1.000 MHz.

A radio receiver has an antenna that intercepts the radiation from the transmitter, as well as the radiation from hundreds of other transmitters sending out signals at other frequencies. At the outset all the transmissions are received together, and it appears to be a hopeless jumble. To solve this problem, the first stage of radio reception is a tuned circuit that precisely selects a desired station. If the tuned circuit is tuned to frequency $f_1 = 1.000$ MHz, then the receiver will pick up Station X. Up to this point, however, no information is being communicated.

To communicate, one needs to make the radio signal audible. To do this, the receiver has a radio-frequency generator of its own (but much less power than the transmitter). Suppose that it generates a signal with a frequency of $f_2 = 1.001$ MHz, and suppose that this is added to the signal from the tuned circuit in a nonlinear device. The nonlinear distortion will create an electrical difference signal having a frequency $f_2 - f_1$, namely 1,000 Hz. This signal is in the audible range and can be converted to a audible signal with a loudspeaker. Out of the speaker comes a 1,000-Hz tone. Wow! Now it is possible to hear radio station X. However, there is still no information being communicated.

Suppose now that the person who is operating the transmitter turns his signal on and off. He might use a telegraph key to do this switching. He can turn it on for a long interval (a "dash") or for a short interval (a "dot"). With a series of dashes and dots the operator can send the letters of the alphabet using Morse code.

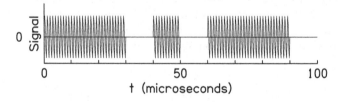

Fig. 21.1 A transmitted 1-MHz radio signal of the type called "continuous wave (CW)" can send Morse code. Shown here is a *dash*, a *dot*, and *another dash*. That makes the letter "K." As shown here, the transmission of *dots* and *dashes* is extremely rapid so that you can see the individual cycles of the carrier

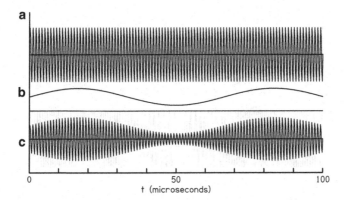

Fig. 21.2 The evolution of an AM broadcast. All the *signals* shown in the figure are voltages, and the *vertical axes* represent volts. Part (**a**) shows a 1-MHz sine wave carrier. In 100 μs there are 100 cycles. When this wave is radiated it will become the electromagnetic radiation of Station X, broadcasting at a frequency of 1 MHz. Part (**b**) shows another electrical wave; it contains the audio information that we desire to transmit using the radio wave. This signal is an audio signal, with a frequency of 15,000 Hz, and it was originally generated from an audible source by the use of a microphone. Part (**c**) shows how the audio signal is transmitted. It is used to modulate the amplitude of the 1-MHz signal. In this way the audible information is incorporated. The process is known as amplitude modulation

At the receiver, the 1,000-Hz tone comes on only when the transmitter is on. If the person at the receiver knows the code, he can decipher what is sent. Now, finally, information can be communicated. This form of broadcasting, called "continuous wave," is the most primitive form. It is illustrated in Fig. 21.1.

21.2 Amplitude Modulation

Modulation makes it possible to transmit more detailed information than simple *on/off* information. With modulation one can transmit speech or music as audio. The concept of modulation begins with the old continuous-wave signal. Quite simply, some aspect of that continuous wave is varied, i.e. modulated, according to the waveform of the speech or music. In amplitude modulation (AM), it is the amplitude of the radio sine wave that is modulated. For instance, Fig. 21.2 modulates a carrier with a 15,000-Hz sine tone as the audio signal.

At the receiving end, the modulation impressed on the signal by the transmitter can be extracted by the AM radio receiver, and the modulation is, of course, the original 15,000-Hz tone that was broadcast. It should now be evident why the 1-MHz radiation is called the "carrier." It acts as a vehicle that *carries* the audio. The frequency of a radio or TV station is the frequency of its carrier. At the receiving end, this carrier is filtered out and only the audio signal is left.

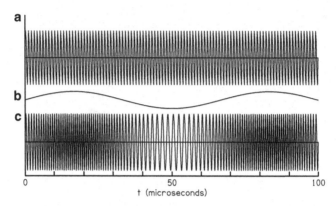

Fig. 21.3 Frequency modulation of a 1-MHz carrier by a 15,000 Hz audio signal. Parts (**a**) and (**b**) are the same as for AM in Fig. 21.2, but the form of the modulation is different

21.3   Frequency Modulation

Figure 21.3 shows a frequency modulated (FM) signal. The FM technique is another means of transmitting an audio signal using radio waves. Now the amplitude is constant. What is modulated is the carrier frequency itself. The carrier frequency is centered on 1 MHz, but the modulation gives this carrier small excursions above and below 1 MHz. The FM receiver has to be different from an AM receiver. Whereas the AM receiver takes variations in the carrier amplitude and converts them to an audible signal, the FM receiver converts carrier *frequency* variations into an audible signal. Receiving information encoded by FM turns out to be less sensitive to static (interference caused by electrical storms); FM provides a better communications channel.

21.4   Bandwidth

The figures above have shown AM and FM signals as functions of time. This time representation is the easiest way to visualize how the modulation affects the carrier. Additional important insight comes from looking at AM and FM from a spectral point of view. At first glance, it would seem that the spectral point of view is very simple. There is a sine wave carrier at 1-MHz. The sine wave has a single spectral component, and so it seems as though the spectrum of the signal would be a single line at a frequency of 1,000,000 Hz. That is actually correct for the continuous-wave signal. However, when a signal is modulated, additional frequencies are added. The additional frequencies are called "sidebands" because

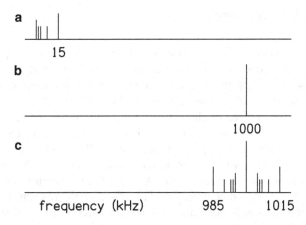

Fig. 21.4 (**a**) An audio signal with components at 5, 6, 7, 10, and 15 kHz. (**b**) A radio frequency carrier with a frequency of 1 MHz (1,000 kHz). (**c**) An AM signal—the 1-MHz carrier is amplitude modulated by the audio signal. Because the audio signal has a frequency as high as 15 kHz, the modulated signal in part (**c**) extends from 985 to 1,015 kHz. Therefore, the bandwidth of the signal is 30 kHz

they appear to the left- and right-hand sides of the carrier frequency. Specifically, when a 1-MHz carrier is modulated by a 15,000-Hz signal there is a component at 1 MHz + 15,000 Hz (the upper sideband) and a component at 1 MHz − 15,000 Hz (the lower sideband).

Of course, the 15,000-Hz tone is a very simple modulation. In practice we would like to broadcast audio with all frequencies from 0 to 15,000 Hz. Therefore, instead of having single discrete sideband components, we get a smeared out spectral region, extending plus and minus 15,000 Hz, either side of the carrier frequency. The total bandwidth is, therefore, 30,000 Hz as shown in Fig. 21.4.

If we wanted to transmit a television signal the bandwidth would have to be much greater. That is because the TV signal includes both the audio and the video information. More information requires greater bandwidth. A conventional analog TV signal has a bandwidth of 6 MHz. Obviously it would not be possible to transmit a TV signal using a carrier of 1 MHz. The sidebands would extend to negative frequencies, and this would greatly distort the signal. Therefore, TV stations broadcast with carrier frequencies that are high, about 100 MHz for VHF, and 500–800 MHz for UHF. The broad bandwidth required for analog TV is the reason that governments worldwide are encouraging digital TV. In fact, they are insisting on it because the information in TV signals can be compressed in the digital domain by eliminating redundancy. Less redundant information means that signals can have a narrower bandwidth. Narrower bandwidth, in turn, means that more TV signals can be squeezed into any given range of frequencies.

21.5   Carrier Frequencies

Worldwide, the transmission of radio signals is regulated. There are specific rules about which radio frequencies may be used, and where, and when. It has to be that way. Otherwise, signals from competing stations would pile on top of one another and communication would be chaotic. It is rather like cars on the street. Only one car is allowed to occupy a given patch of street at any given time. Particular frequency bands have been assigned to particular kinds of broadcasting. In the USA, AM broadcasting is done with carriers in the range from 550 to 1,600 kHz (or 1.6 MHz). FM broadcasting is done with carriers in the range from 98 to 106 MHz. Because the AM range is so narrow (only about 1 MHz) AM radio stations are limited in bandwidth. In fact, it would be illegal for an AM station to broadcast a signal with a bandwidth of 30 kHz as in the example above. However, an FM station is allowed to broadcast a signal that wide.

Analog TV broadcasting takes place in several frequency ranges. Channels 2–6 are found between 54 and 88 MHz. Channels 7–13 are found between 174 and 216 MHz, and channels 14–69 are found between 470 and 806 MHz.

Exercises

Exercise 1, Frequency and wavelength revisited

(a) AM radio broadcasts at frequencies of about 1 MHz, i.e. 10^6 Hz. Show that the wavelength is about 300 m. (b) Microwaves have frequencies as high as 10^{11} Hz. Show that the wavelength is 3 mm.

Exercise 2, AM radio

Draw the spectrum of a 1-MHz carrier amplitude modulated by a 15,000-Hz audio signal.

Exercise 3, The bandwidth cost of video

How many times greater is the bandwidth of an analog TV signal compared to the 30,000-Hz bandwidth of an FM signal?

Exercise 4, FM vs AM

Why do you suppose that FM transmission is less susceptible to static than AM transmission? Consider static to be a noise that is added to the electromagnetic radiation on its way to the receiver.

Exercise 5, Who hears it first?

The back row of the opera house is 150 ft away from a singer on stage. The broadcast microphone may only be 2 ft away. Who hears the singer first? Is it the person in the audience sitting in the back row at the opera, or is it you, sitting in the hot tub perhaps several thousand miles away? [Hint: Remember that the broadcast comes to you at the speed of a radio wave, namely the speed of light.]

Exercise 6, Bandwidth, information and economics

What is the connection between bandwidth and the amount of information that is transmitted? Is bandwidth scarce? Must it be regulated?

Exercise 7, Static

The antenna of a radio receiver not only picks up all the radio transmitters within range, but it also picks up noise (called "static" in radio communications terminology) from electrical storms all around the world. The input tuning of the receiver helps reduce this noise. How do you think that this works?

Exercise 8, TV broadcasting

The text says that TV signals have a bandwidth of 6 MHz. (These were the USA standard—NTSC—signals prior to digital TV.) Is this bandwidth consistent with the assignment of frequency ranges for TV channels given in the section called "Carrier frequencies?"

♠

Chapter 22
Speech

Speech is our most basic and most important means of human communication. Speech conveys more than mere words. It conveys shades of meaning, emotion, attitudes, and opinions, sometimes even more completely than we as talkers would like. As a topic for study, human speech has endless fascination for acousticians, phoneticians, linguists, and philosophers. This chapter can only scratch the surface of this vast topic. It is mainly devoted to the *production* of speech by human talkers. It introduces the anatomy of the vocal tract and then describes how the components of the vocal tract function to create speech sounds, particularly vowels—the basis of singing.

To an unusual extent, the goal of this chapter is merely to cause the reader to become aware of things that he or she already knows from years of experience in talking. Properly read, this chapter requires some vocal involvement from the reader, if only to check that certain sounds are really made the way that the text says that they are made. If it's done right, it should be a noisy experience.

22.1 Vocal Anatomy

Speech sounds are made by using air from the lungs. The lungs provide all of the energy that ultimately ends up in a speech waveform. The air from the lungs passes through the trachea, to the larynx—including the vocal folds, and then to a part of the vocal tract that serves as a resonator. We consider those processes in turn (Fig. 22.1).

Diaphragm and lungs: The diaphragm is an arched sheet of muscle separating the lung cavity from the abdominal cavity. To understand what happens as you inhale, take a few deep breaths and use your hands to feel how your upper body moves.

W.M. Hartmann, *Principles of Musical Acoustics*, Undergraduate Lecture Notes in Physics, DOI 10.1007/978-1-4614-6786-1_22,
© Springer Science+Business Media New York 2013

Fig. 22.1 Vocal anatomy, cross section after Farnsworth, 1940

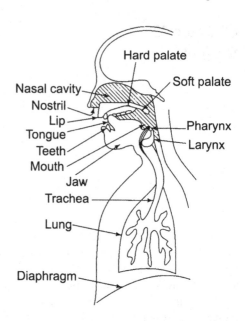

Notice that when you inhale, your belly seems to protrude as the diaphragm is lowered to expand the lungs, and in the process moves other organs around lower down. Notice too that your upper chest walls rise. Both actions expand the chest cavity, allowing air to rush in because of normal atmospheric pressure. This process of inhalation, or inspiration, can expand the volume of air in the lungs from 2 to 7 L for the average adult human. However, in normal quiet breathing, the change in lung volume is only 10% of that.

Expiration (or exhalation) occurs when the normal elasticity of the lungs and chest walls forces air out of the lungs. Also, the lungs have a reserve capacity that can be made to expel additional air using expiratory muscles. When we stress certain syllables while speaking, the momentary added air volume is expelled by chest-wall muscles called "intercostal muscles."

A liter

The *liter* is a metric unit of volume. It is equivalent to 0.001 cubic meters or 1,000 cubic centimeters (cc). English equivalents to a liter are 61 in.³ or slightly more than a USA quart.

Trachea: The trachea, commonly known as the "windpipe," allows air to flow in and out of the lungs. As the trachea descends toward the lungs, it divides into two bronchi (bronchial tubes), one for each lung. The trachea has a mucous lining that helps to catch dust particles, preventing them from entering the lungs. In that sense, its function is the same as the air intake and air filter on your car's engine.

Larynx: At the top of the trachea, and contiguous with it, is the larynx or voice box. The larynx serves many functions, not the least of which is to channel air to the trachea and food to the esophagus. Within the larynx is the *glottis*, consisting of the vocal folds (vocal cords)—two mucous membranes with a separation that is under exquisite, active, conscious control by the brain.

As you exhale normally through the trachea, the vocal folds are apart and air passes by them. If you hold your breath, the vocal folds are tight shut. When you make a *voiced* sound, the vocal folds are loosely together. The vocal folds vibrate against one another to make a buzz driven by a steady stream of air from the trachea. For unvoiced sounds the vocal folds are apart, as in normal breathing. Because there is no vocalization, unvoiced sounds have to be made by creating constrictions using pharynx, soft palate (vellum), tongue, teeth, and lips and then blowing air past these constrictions. The constrictions may be fixed in time, as for the sound "ssss," or the constrictions may suddenly open as for the sound of "p." In whispering, the vocal folds are somewhat apart, but they vibrate in a noisy way to create a sound that can be filtered later, just as for voiced sounds.

Resonators: The pharynx (pharyngeal cavity), mouth, and nasal cavities have resonances that act as filters, reinforcing particular frequency regions of the voiced or whispered sounds from the larynx. The frequencies of the resonances can be precisely controlled because of all the flexibility that you have with the shape of your mouth cavity and your tongue and lips. The different resonances are responsible for creating the different vowel sounds.

22.2 Voiced Sounds

The buzzing sound made by vibrating vocal folds is normally a periodic wave, and it has a complex spectrum with many harmonics. The vocal buzz, waveform and spectrum, is shown in Fig. 22.2. To some degree, the vocal buzz resembles the sounds of wind musical instruments—the brass instruments and woodwinds. Both the vocalized sounds and the wind instrument sounds begin with air from the lungs. Both sounds are generated by some physical element that vibrates with periodic motion in the air stream, as controlled by the intelligence of the person producing the sound. The analogy is particularly valid in the case of singing where sustained sounds have stable frequencies or frequencies that move only slowly or regularly. Physically, however, there is an important difference between the production of voiced sounds and the production of wind instrument sounds.

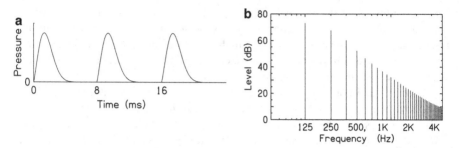

Fig. 22.2 (**a**) The waveform of a vocal sound measured in the throat shows peaks of air pressure when the vocal folds are open and intervals of zero pressure when the vocal folds are closed. The waveform is periodic with a period of about 8 ms, typical of an adult male voice. (**b**) The spectrum shows a fundamental frequency of 125 Hz (=1,000/8) and strong harmonics. Different shapes of oral cavities further upstream will lead to different vowel sounds. Note that in this figure, both axes are logarithmic. The harmonics are really equally spaced by 125 Hz, as expected for a periodic tone, but the logarithmic horizontal axis (a scale of octaves) does not give that impression

The vibrating element in the wind instrument, for example a trumpet player's lips or a clarinet player's reed, requires feedback from the rest of the instrument in order to produce the tone with the intended pitch. It is not so with voiced sounds. Although the parts of the vocal tract that follow the larynx (pharynx, mouth cavity, and nasal cavity) act as resonators or filters, they do not provide important feedback to the larynx. The larynx is capable of generating the voiced tone, with the correct frequency, all by itself without help from the rest of the vocal tract. Consequently, the human voice is easier to understand physically than are wind musical instruments. That is why the chapter on the human voice comes first in this book, before the chapters on musical instruments.

The most prominent voiced sounds are the vowels. Vowel sounds carry most of the energy in speech, though not necessarily most of the information. Because of their power, vowels are the main element in singing. It is easy to sing without consonant sounds. It is impossible to sing without vowels.

To make vowel sounds, one shapes the vocal tract—mouth and nasal cavities— to give these cavities particular acoustical resonances as shown in Fig. 22.3. These resonances cause particular harmonics of the vocal fold buzz to be emphasized. Thus, the vocal folds create a complex spectrum and the vocal tract filters it. The frequency regions containing strong harmonics due to vocal tract resonances are called "formants." It is usual to characterize vowel sounds by three formants, though it is possible to find as many as five.

The formant concept

The idea of a formant is nothing new. A formant is a band of frequencies that is prominent in the output of an acoustical source. This band of frequencies is prominent because the source includes a filter, or resonator, that emphasizes those frequencies.

The concept of a formant can be illustrated in a speech or music context in which the band of strong frequencies remains constant while the fundamental frequency of the sound changes. For instance, if a singer sings three different notes all with the vowel "AH" there are three different fundamentals, but the range of emphasized frequencies stays the same. A formant near 1,000 Hz strongly emphasizes the eighth harmonic of a 125-Hz tone, or the fifth harmonic of a 200-Hz tone, or the third harmonic of a 333-Hz tone.

A mute inserted into the bell of a brass instruments (e.g., trumpet or trombone) introduces a formant into the sounds of the instrument. This changes the tone color of every note played. For instance, a trumpet mute called a "cup mute" introduces a resonance peak extending from 800 to 1,200 Hz. Although the trumpet may play dozens of different notes, each with its own fundamental frequency and harmonic series, the formant, 800–1,200 Hz, applies equally to every note and leads to an unmistakable coloration of the sound.

Formant frequencies for different American vowel sounds are given in Table 22.1. The table shows the average formant frequencies for 76 talkers—33 men, 28 women, and 15 children. Average formants for adult women are higher than those for adult men because, on the average, vocal tracts are shorter for women. If you think of a vocal tract as a cylindrical pipe open at one end and closed at the other (Chap. 8), you are not surprised to learn that a shorter pipe has higher resonance frequencies. Children have much shorter vocal tracts and correspondingly much higher formant frequencies. As listeners, we associate high formant frequencies with small creatures. We also associate high fundamental frequencies with small creatures. The high fundamental frequencies of children occur because their vocal folds are lighter than for adults. Thus, high formant frequencies and high fundamental frequencies occur for very different reasons.

22.3 Speech Sounds

Years ago, you learned that speech is composed of vowels and consonants. This distinction is not necessarily wrong, but it greatly oversimplifies the phonological and linguistic nature of speech and language. Although it is not hard to define a

Table 22.1 Average formant frequencies and levels for vowels spoken by mainly American talkers

Vowel	EE	I	E	A	AH	AW	U	OO	UH
(as in)	heed	hid	head	had	hod	hawed	hood	who'd	hud

Frequencies (Hz)

F1

	EE	I	E	A	AH	AW	U	OO	UH
Men	270	390	530	660	730	570	440	300	640
Women	310	430	610	860	850	590	470	370	760
Children	370	530	690	1,010	1,030	680	560	430	850

F2

	EE	I	E	A	AH	AW	U	OO	UH
Men	2,290	1,990	1,840	1,720	1,090	840	1,020	870	1,190
Women	2,790	2,480	2,330	2,050	1,220	920	1,160	950	1,400
Children	3,200	2,730	2,610	2,320	1,370	1,060	1,410	1,170	1,590

F3

	EE	I	E	A	AH	AW	U	OO	UH
Men	3,010	2,550	2,480	2,410	2,440	2,410	2,240	2,240	2,390
Women	3,310	3,070	2,990	2,850	2,810	2,710	2,680	2,670	2,780
Children	3,730	3,600	3,570	3,320	3,170	3,180	3,310	3,260	3,360

Relative levels (dB)

	EE	I	E	A	AH	AW	U	OO	UH
L1	−4	−3	−2	−1	−1	0	−1	−3	−1
L2	−24	−23	−17	−12	−5	−7	−12	−19	−10
L3	−28	−27	−24	−22	−28	−34	−34	−43	−27

steady vowel, as we have done with the formant concept, the variety of human speech behavior makes even the study of vowel sounds a complicated one. A thorough approach to a more advanced treatment is beyond the scope of this book. This section on speech sounds will deal with the some classes of speech sounds with emphasis on the means of production.

diphthongs: Within the vowel context there are diphthongs, combinations of two vowels that are interpreted as a single speech sound. The word "toil" is an example. As for the word "out" from the speech spectrogram, it is impossible to pronounce the vowel sound in "toil" without changing the shape of the vocal tract. Even the pronunciation of the first letter of our English alphabet seems to end in an EEE sound if you listen to it closely.

glides: A glide is a vowel-like sound that requires the vocal tract to move. The WH sound in "when" is an example.

Fig. 22.3 Creating a voiced vowel. (**a**) The spectrum of a vocal sound measured in the throat from Fig. 22.2. (**b**) The gain of the vocal tract filter making the vowel E as in "head." (**c**) The spectrum of the vowel finally produced by the talker. The peak near 500 Hz is the first formant. It appears to be broader than the other peaks, but that is a distortion of the logarithmic plot. In units of linear frequency, the first formant peak is actually the narrowest

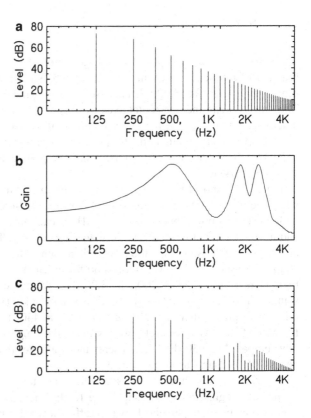

fricatives: Fricatives are consonant sounds that are essentially noise. They are made by creating a constriction somewhere in the vocal tract and blowing air past the constriction to make a turbulent flow with a noisy character. The FF sound is made with a constriction between the upper teeth and lower lip. To make the TH sound you make a constriction with your tongue in back of your teeth. The HH, as in "hah" comes from a constriction in the glottis itself. The sounds known as sibilants are the hissy fricatives like SS and SH. Spectrally, fricative sounds are broadband with no harmonic structure. However, different fricatives have different frequency regions of spectral strength. For instance, the SS sound has much greater high-frequency power than the SH sound.

plosives: As the name implies, a plosive is a little explosion. Plosives (sometimes called "stops") are made by building up air pressure behind a closed constriction and then releasing it. The plosives P,T, and K are not voiced. The plosives B,D, and G are normally accompanied by voicing. Plosives, as well as fricatives, are said to be manners of articulating consonant sounds.

22.4 Spectrograms

A spectrogram, for example Fig. 22.4, is a plot of power in a vocalization as a function of frequency and time. The frequency is plotted on the vertical axis and runs from 0 to 4,000 or 5,000 Hz because this is the vital frequency range for speech. The time is along the horizontal axis, as for an oscilloscope, but the time scale is much longer than the time for a trace on the oscilloscope. From left to right the time is several seconds—enough time to speak an entire sentence. The power that occurs at a particular frequency (vertical axis) and at a particular time (horizontal axis) is shown by darkness.

Figure 22.4 shows a spectrogram for the utterance, "Joe took father's shoe bench out." The first sound that occurs is the "J" sound in "Joe." Notice that it is a noise with most of its power between 2 and 4 kHz. Next comes the "OH" sound in "Joe," and it shows motion in time. The formants are moving. There is some weak noise energy for the "t" sound in "took," and then comes the short "oo" vowel sound in "took." The short "oo" vowel is too short for much motion. It appears to show dark regions near 400, 1,300, 2,200, and 3,200 Hz. We can compare that with the "U" sound in Table 22.1, as in "hood." If this is a male talker (we don't really know from this spectrogram), then formants are expected at 440, 1,020, and 2,240 Hz. These first three formants agree pretty well with the dark regions, though 1,020 Hz seems low compared to 1,300 Hz. This discrepancy indicates that individual talkers aren't necessarily the same as the average talker.

Figure 22.4 shows that the "s" sound that ends the word "father's" and the "sh" sound that begins the word "shoe" are totally merged in time. The breaks that experienced listeners perceive between different words may have no real acoustical

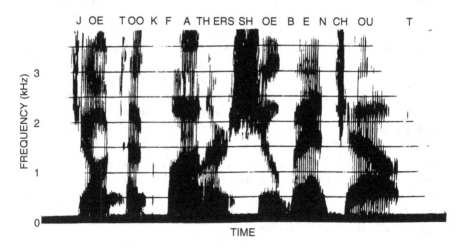

Fig. 22.4 Speech spectrogram. Power is indicated by *dark regions*. From W.J. Strong, J. Acoust. Soc. Am. **41**, 1434 (1967)

existence in the speech waveform. The segmentation of different words from the continuum can be a problem for listeners who are less familiar with the language.

Something is moving a lot in the "oe" sound that ends the word "shoe." Perhaps it is a second formant that is moving. In any case, there is no necessary motion in pronouncing the word "shoe." You will find it possible, though perhaps a little unnatural, to say the word "shoe" with all your articulators in a fixed position. Individual modes of speaking show greater or lesser motion. By contrast, it is impossible to pronounce the word "out" without a lot of motion of the mouth. That can be seen in the huge time variation in the first and second formants at the end of the sentence in Fig. 22.4. It looks as though the second formant drops from about 1,500 to 1,000 Hz.

The speech spectrogram is a very useful method of studying speech. It is detailed, but not too detailed. Spectrograms of different individuals saying the same sentence can show what parts of speech are essential to convey the intended words and what parts are individual differences. The spectrogram gives clues as to the articulatory gestures used in speech, but they are only cues because there is more than one way to make any given speech sound.

Exercises

Exercise 1, Breathing-in speech.
Prove that it is not necessary to breathe out when you speak. It is possible to speak while breathing in. Can you learn to control this kind of speech?

Exercise 2, We are all windbags!
The text says that the lung capacity of the average human is about 5 L. A table of weights and measures says that 1 L is 0.2642 gallons (that's 0.22 British imperial gallons). (a) Calculate the average lung capacity in gallons; (b) in quarts.

Exercise 3, Feeling the buzz
(a) Feel the buzz of your vocal folds when you make a voiced sound. (b) Which of the following sounds are voiced: FF, SS, B, V, P, T?

Exercise 4, The nose knows.
(a) Which speech sounds can you make with your mouth closed? (b) Practice making the transition between "M" and "N" with your mouth closed and in one position. Can you get a friend to distinguish between these two sounds?

Exercise 5, Diphthongs or not?
Say the vowels as you learned them in school: A, E, I, O, and U. Which vowels would you expect to sound most nearly the same if played backwards?

Exercise 6, Plosive sounds
Where are the "explosions" taking place in your mouth in the sequence, "Pah, Tah, Kah, Gah?" Does this sequence go from front to back or from back to front?

Exercise 7, Fricative sounds

Fricative sounds are made by blowing air past a constriction. Where are the constrictions in the sequence, "Sah, Fah, Thah, Hah?"

Exercise 8, Whispered vowels

Say a vowel out loud. Now whisper the same vowel. Did you have to change the shape of your vocal tract when you started to whisper? Why?

Exercise 9, Chimeras

The discussion of Table 22.1 noted that children have high fundamental frequencies because their vocal folds are light, and they have high formant frequencies because their vocal tracts, from the pharynx on up, are small. Can you imagine a creature with low fundamental frequencies and high formants or vice versa?

Exercise 10, Read a spectrogram

In the spectrogram, identify the spectral character of vowel formants, formant transitions, fricatives (like sibilants), and plosives.

Exercise 11, The remarkable "AH"

Refer to Table 22.1 and notice the separation in frequency between the first and second formants of the AH and AW vowel sounds as in "hod" and "hawed." Notice that the separation is the smallest separation for any of the formants of any of the vowels in the table. Figure 22.4 has an AH sound in "father." Can you now account for the big blob of power below 1,500 Hz in the spectrogram of that word?

Exercise 12, Benched!

Look at the blobs of power in the "E" sound in the word "bench" in Fig. 22.4. What are the frequencies for these regions? Do they agree with the formant frequencies in Table 22.1 for the "E" sound as in "head?"

♠

Chapter 23
Brass Musical Instruments

The family of musical instruments known as brass instruments includes the trumpet, trombone, French horn, and tuba, among others. The ranges of these well-known orchestral instruments are given in Appendix C. If you guessed that brass instruments are made out of brass you would be right, but they are sometimes made out of other materials as well. For instance, the vuvuzela is made out of vinyl. What really characterizes the brass family, as we define it, is that the sound originates with the vibration of the player's lips in a mouthpiece. Some authors refer to brass instruments as "lip-reed" instruments. Another characteristic of the brasses is that they all have a bell, a flaring end of the horn that radiates *all* the sound.

23.1 Sustained-Tone Instruments

Brass instruments are sustained-tone instruments, like the bowed strings, the woodwinds, and the human singing voice. All these instruments make tones that carry melodies and can be used in close harmony. These tones are approximately periodic, though they do not have the perfect cycle-to-cycle redundancy of the periodic complex tones from an electronic function generator. They include intentional expressive variations such as glides and vibrato (frequency modulation) and tremolo (amplitude modulation), as well as unintended noise and glitches. These tones might be called "adequately periodic," in the sense that the perceived attributes of harmonic fusion and unambiguous pitch that apply to the ideal tones from precision electronic function generators also apply to them.

Sustained tones, like those from brass instruments, are close enough to periodic that the standard physical description of them begins with a perfectly periodic prototype and then deals with variations as deviations from that prototype. The prototype is not just an abstract idealization. With special care it is possible to produce a sustained tone with a brass instrument that is stable on an oscilloscope screen for many seconds. That means that the relative phases among the harmonics do not

W.M. Hartmann, *Principles of Musical Acoustics*, Undergraduate Lecture Notes in Physics, DOI 10.1007/978-1-4614-6786-1_23,
© Springer Science+Business Media New York 2013

change over the course of many seconds. That, in turn, means that the harmonic frequencies deviate from integer multiples of a fundamental frequency by much less than 1 Hz. Thus, if brass-instrument tones differ at all from perfect periodicity, it is because the performer makes it so. The basic mechanism of sound production is capable of making ideally periodic tones.

Harmonic Partials from Inharmonic Resonances The harmonics of a brass instrument tone (essentially perfect)should not be confused with the resonances of the instrument itself. Brass instruments are hollow tubes that have been fashioned to make air columns with modes of vibration with frequencies (resonance frequencies) that are approximately in a harmonic relationship. These modes, or resonances, are responsible for generating the harmonics of tones, but they may easily deviate from perfect harmonicity by 5 or 10 %. How a musical instrument generates a tone with partials that are in a perfect harmonic relationship from resonances that are only approximately harmonic is an essential part of the nonlinear tone generation process in wind instruments.

Thus, there are two essential problems in the understanding of brass instruments. The first is to understand the resonances of the instrument. The second is to understand how these resonances can be made to produce tones with clear pitches and stable harmonic spectra. We deal with the resonance problem first.

23.2 Evolution of the Resonances of a Trumpet

Any brass-instrument player can pick up a piece of copper pipe and play a few notes by buzzing his lips in one end of the pipe. If the pipe is 1 m long, the notes will have fundamental frequencies of about 260, 430, and 600 Hz. These frequencies fit a pattern; they are approximately 3×86, 5×86, and 7×86. Returning to the study of cylindrical pipes in Chap. 8, you discover that this pattern looks like a pipe that is open at one end and closed at the other. The closed end is closed by the player's lips. You can calculate the base frequency to be $v/4L$ or $344/4 \cdot 1 = 86$ Hz, and that agrees with the observed result. You also expect there to be only odd resonances. It all works out, except that apparently the brass instrument player is unable to make a tone from the lowest resonance at 86 Hz. However, if a clarinet player comes along and puts her reed mouthpiece on the end of the pipe, she can play a tone corresponding to that low-frequency resonance.

The Bugle The first step in developing a useful instrument is to add a mouthpiece and bell to the cylindrical tube as shown in Fig. 23.1. The function of the mouthpiece is mainly to provide a comfortable and convenient place to vibrate lips. The bell has two functions. First, it helps to radiate the tone, especially the high harmonics of the tone so that the brass instrument sounds bright. Second, the bell greatly modifies the resonance frequencies of the instrument. They are no longer odd integer multiples of a base.

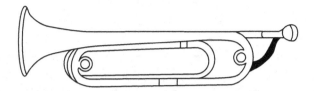

Fig. 23.1 The bugle is a cylindrical pipe for most of its length. It has a mouthpiece on one end and a bell on the other. The cylindrical pipe is bent around to make the bugle easy to carry, but all those bends have little effect on the playing of the instrument. If the bends were all straightened out to make a long pipe the bugle would not sound any different

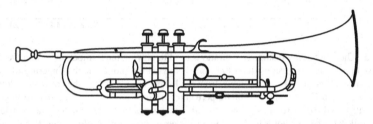

Fig. 23.2 The trumpet has three valves. When the piston is depressed in the valve an extra length of tubing is added to the horn

If we give the brass-instrument player a bugle, he can play many more notes than with the simple pipe. He can play notes with fundamental frequencies of approximately 196, 293, 392, 493, 587, 698, and 784 Hz. These frequencies fit the pattern 2×98, 3×98, 4×98, 5×98, 6×98, 7×98, and 8×98. Appendix E shows that these fundamental frequencies correspond to the notes *G3, D4, G4, B4, D5, F5, G5*. The first five of these notes are used in traditional bugle calls, like taps or reveille. The important point is that the addition of a flaring bell to the cylindrical pipe has deformed the pattern of resonances to look like successive integers (2, 3, 4, ...) not like odd integers only. You will notice that the integer 1 is missing from the list. The lowest-frequency mode of a horn with a bell has a frequency too low to fit the pattern. It is not used musically.

Adding the Valves Like half a dozen other brass instruments, the trumpet (Fig. 23.2) has three valves, numbered 1, 2, and 3, with valve number 1 being closest to the player. Each valve is fitted with a piston. In normal playing, a piston is either up or down. Intermediate positions are not used. When the player depresses a piston, an extra length of cylindrical tubing is inserted into the cylindrical section of the instrument. Piston 1 inserts about 18 cm. Piston 2 inserts about 10 cm. Piston 3 inserts about 30 cm.

Figure 23.3 shows how a player uses the valves to play notes in between the standard notes of the bugle. The figure shows that the trumpet is essentially seven bugles. The first bugle is made with the open horn, and it can play the trumpet's notes *C, G, C, E, G* ... The second bugle is made by depressing piston 2, and it is tuned half a tone lower, playing *B, F♯, B, D♯, F♯* ...

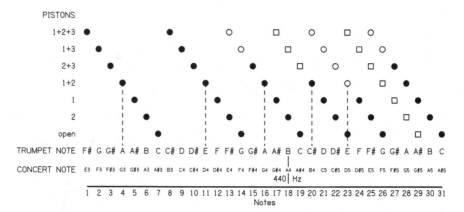

Fig. 23.3 There are seven piston configurations used in playing the trumpet, including the "open-horn" configuration—no pistons depressed. *Solid dots* on the plot show the fingering (depressed piston configuration) for 31 trumpet notes from the lowest written note to the note called "high C." (Notes above high C are used by athletic performers.) *Open circles* show alternative fingerings, not normally used. *Open squares* show alternative fingerings that lead to notes that are badly out of tune—all flat. The five notes used in bugle calls are indicated by *dashed lines*. Labels near the bottom of the plot show the names of the musical notes. So called, "accidentals"—sharps (♯) and flats (♭)—are here indicated by sharps only. In fact, A♯ is equivalent to B♭; C♯ is equivalent to D♭, etc. The trumpet is a transposing instrument, and its nomenclature for musical notes does not agree with that for the piano or violin, the basis of "concert pitch." Because the trumpet is a B-flat instrument, the trumpet player's "C" corresponds to B♭ (A♯) on the piano or violin

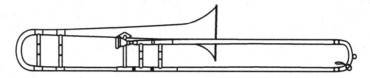

Fig. 23.4 The trombone has a slide that can be varied continuously, but there are seven distinct positions that correspond to notes of the scale

Other valved brass instruments, cornet, baritone horn, and tuba have very similar fingerings. Although it does not look like it, a trombone (Fig. 23.4) is essentially a trumpet, tuned an octave lower because the pipe is twice as long. Every trombone player learns seven positions for the slide—equivalent to the seven valve configurations on the trumpet.

23.3 Tone Production: Feedback and Nonlinearities

To a casual observer brass instruments look complicated. They look like a plumber's nightmare of valves, slides, crooks, and tubing bent into intricate shapes. (This is especially true of the French horn (Fig. 23.5), which has a great length of tubing,

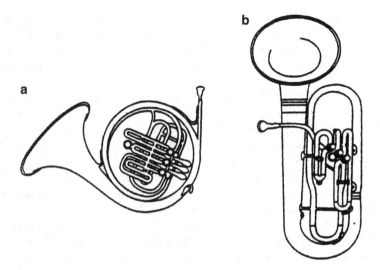

Fig. 23.5 (a) The French horn and (b) the baritone horn

all curled up into a convenient package.) But from a physical point of view, what is truly complicated about a brass instrument is the interaction of the player's lips—buzzing in the mouthpiece—and the resonances of the plumbing. This section tries to deal with that interaction.

The mouthpiece of a brass instrument can be disconnected from the rest of the instrument. If you give a bugle player the mouthpiece alone, the player can vibrate his lips without difficulty in the mouthpiece to produce a buzzy sound, rather like an obnoxious duck. While this buzzy sound continues, the rest of the instrument can be slid onto the mouthpiece to produce a normal bugle tone. What is interesting about this little experiment is the sensation experienced by the player. With the mouthpiece alone, the lip vibration feels chaotic. As the rest of the instrument is added, the lips are forced into a more orderly pattern of vibration and the player can feel this happening.

The cause of this change can be described as *feedback* from the horn to the mouthpiece, or as *reflections* of puffs of air from the horn back to the mouthpiece, or as the result of *standing waves* in the horn. The standing waves can be called "resonances" of the horn. These descriptions are essentially equivalent. Because of its strong resonances, the horn reinforces lip vibrations near its resonance frequencies. If the horn resonance frequencies are nearly in a harmonic relationship, they cause the lips to vibrate periodically and the puffs of air admitted into the horn by the lips through the mouthpiece are caused to have a well-defined shape in space and time. Well-defined pulses lead to a brassy tone with strong high harmonics. If the resonance frequencies of the horn are not well aligned in a harmonic relationship, the puffs of air are less compact in shape and the tone quality becomes duller.

The bugle player can get at least six different musical notes from the instrument. A trumpet player can get at least six for every configuration of the pistons. Of course, for each configuration the horn has only one set of resonances. Thus, it is evident that the player must be using the resonances in different ways to get all the different notes.

Figure 23.6 shows the resonances of a trumpet with no pistons depressed. You can see 14 resonant peaks there. Their frequencies are 92, 233, 364, 464, 587, 710, 819, 937, 1,065, 1,183, 1,320, 1,451, 1,579, and 1,706 Hz. These are the peaks that can be used to create tones. For example, the lowest tone of the open horn (no pistons down) has a fundamental frequency of 233 Hz. Figure 23.6 includes circles indicating the harmonics of the 233-Hz tone. Their frequencies are 233, 466, 699, 932, 1,165, 1,398, and 1,631 Hz. The first five harmonics are supported by the nearby resonances, with peaks at 233, 464, 710, 937, and 1,183.

If the musician wants to play a different note, still with no pistons depressed, the musician will tighten the lips, and perhaps blow a little harder. The result will be that the resonances are used in a different way to create a tone with a different fundamental frequency. Learning how to control the lips in such a way as to make flexible use of the resonances of the horn is the essence of learning how to play a brass instrument. The resonances that contribute to the different notes will become clear on the completion of Fig. 23.6 in Exercise 23.2.

The collaboration between vibrating lips and the resonances of the horn in creating a tone has several consequences. These are described below in terms of resonances numbered 1–14 starting from the left.

Consequence 1: Harmonic Amplitude vs Resonance Peak Height and Tuning If a resonance of the horn is close in frequency to a harmonic of the tone and the resonance peak is tall, the corresponding harmonic will normally be strong in the tone. For example, in the 233-Hz trumpet tone you can expect the second and third harmonics to be strong because the 4th and 6th resonance peaks in Fig. 23.6 are tall. For the note an octave higher, with fundamental frequency 466 Hz, the fundamental should dominate because the resonance at 464 is stronger than the other contributing resonances as you will see when you do Exercise 23.2. Because these resonances are fixed in frequency, as the playing (fundamental) frequency changes, tall resonances lead to formants, characteristic of the instrument.

Consequence 2: Harmonic Amplitude vs Intensity If the player blows with greater force, the amplitudes of all the harmonics increase. However, the amplitudes of the higher-frequency harmonics increase more than the amplitudes of the lower-frequency harmonics. Therefore, in relative terms, the high harmonics become increasingly important, and the tone color becomes brighter.

Consequence 3: Playing Frequency vs Intensity Each resonance that helps to create a tone plays a role in determining the playing frequency. For example, consider the five resonances that help create the 233-Hz tone. Resonance number 2, with a peak frequency at 233 Hz, wants the playing frequency to be 233 Hz. Resonance 4, with a frequency of 464 Hz, would like to create a second harmonic of a 232-Hz tone, and therefore this resonance would like the playing frequency to be

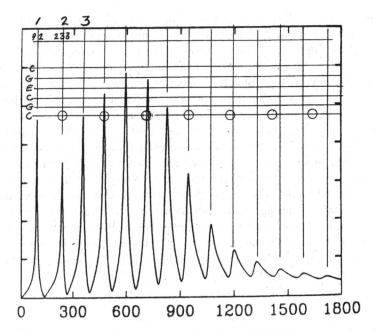

Fig. 23.6 The response *curve* for a trumpet indicates the amplitude of pressure variations at the mouthpiece caused by feedback from the rest of the horn. This feedback will help the player's lips vibrate with frequencies of peaks of the response curve. Peaks are called resonances, the first three are numbered. A few of their frequencies are given too. *Circles* indicate the frequencies of harmonics of low C, with a playing frequency of 233 Hz

232 Hz. That is because $464/2 = 232$. Resonance 6, at 710 Hz wants to create the third harmonic of 237 Hz because $710/3 = 237$. Similarly resonance 8, at 937 Hz wants to create the fourth harmonic of 234 Hz, and resonance 10 at 1,183 Hz wants to create the fifth harmonic of 237 Hz.

Evidently the different resonances are voting for different playing frequencies. But the voting is not equal. The votes are weighted by the contribution to the tone. This weighting leads to small changes in frequency with increased intensity. As described in Consequence 2, the higher harmonics become relatively more important with increasing intensity. Harmonics 3, 4, and 5 would all vote for a playing frequency that is higher than 233 Hz. As they become more important with increased blowing pressure, the playing frequency of this note on this instrument will tend to rise slightly.

Consequence 4: Bending Notes Establishing the playing frequency of a brass instrument can be imagined in three steps. (1) First the musician tightens his lips and sets the blowing pressure to establish the general frequency range of the note to be played. (2) Then the resonances of the horn determine more precisely what the playing frequency shall be, within the range established in step (1). The resonances determine the reflections of puffs of air backwards toward the mouthpiece that cause the musician's lips to open at the right time to admit a new puff of air. If a trumpet player is playing "high C," with a fundamental frequency of 932 Hz, then the player's lips open and close 932 times per second, admitting 932 puffs of air every second. (3) Finally, the musician can establish some micro control of the playing frequency. He can "lip" the frequency up or down. Because the playing frequency is always a compromise among the different resonances, and because the resonance peaks themselves are not infinitely sharp, a musician can tighten or relax his lips ever so slightly to cause the instrument to run a little bit sharp or flat compared to the best combination of resonances. In this way, a musician can bend notes. Bending notes can be used to play better in tune with another instrument or for expressive effect, as in jazz performance. If a trumpet player wants to bend a note by a large frequency difference, he can deliberately wreck the pattern of resonances of his instrument by closing a valve (usually the second valve) half way. When the valve is caused to be neither fully open nor fully closed, the resonances become pale shadows of what they were normally. Because the resonances are then not so strong, the musician's lips have more micro control of the playing frequency, but the tone quality becomes stuffy and not at all brassy.

Exercises

Exercise 1: Trumpet tones
   With an open horn, a trumpet can play notes that the trumpet player calls C_4, G_4, C_5, E_5, G_5, C_6. The trumpet is a "transposing" instrument, which means that the real notes (concert notes) are Bb_3, F_4, Bb_4, D_5, F_5, Bb_5. These notes have fundamental frequencies: 233, 349, 466, 587, 698, 932 Hz, respectively.

(a) From the frequencies, show that all the notes called C are in octave relationships.

(b) Show that the notes called G are a musical fifth above the notes called C because the frequency ratios are about 3/2.

Exercise 2: Trumpet resonances and trumpet tones
   Refer to Fig. 23.6, the large resonance curve for the open trumpet. Do the following steps on that plot. (a) Complete the numbering of resonances, 1–14. (b) Complete the list of peak frequencies on the plot. They are: 92, 233, 364, 464, 587, 710, 819, 937, 1,065, 1,183, 1,320, 1,451, 1,579, and 1,706 Hz. (c) Use circles to indicate which resonances are used to make the six tones with the fundamental frequencies given in Exercise 1 (notes are C, G, C, E, G, C in trumpet notation). The first tone, C 233-Hz, has been done for illustration.

Exercise 3: Open or closed or neither?
   A bugle resembles a pipe that is open at one end (the bell) and closed at the other (the mouthpiece). However, in an important sense it better resembles a pipe that is open at both ends. In what sense is that? How do you account for this apparent paradox?

Exercise 4: Trumpet valves and trumpet tones

A trumpet has three valves (so do many other brass instruments). In normal playing, a valve is either open or closed. Show that the valve combinations can be represented as a 3-bit word. Therefore, show that there are eight possible combinations. Which combination is missing from Fig. 23.3?

Exercise 5: Trombone and trumpet

The trombone tubing is about twice as long as the trumpet. Show why you expect it to sound an octave lower.

Exercise 6: French horn and trumpet

The French horn tubing is about twice as long as the trumpet, but the playing frequencies are about the same as the trumpet. How can this be? [Hint: The notes that can be played in a single valve position on a French horn are much closer together than the notes that can be played on a trumpet.]

Exercise 7: The abused bugle

The author owns a bugle that was thoroughly dented during several years of camp. The bugle still plays the right notes (correct fundamental frequencies) but the tone color is bad. Can you explain?

Exercise 8: Speedy!

The text says that when a trumpet player plays "high C," his lips open and close 932 times per second. Bozo disagrees. He says that nobody can move his muscles, even lip muscles, at such a rapid rate. Set Bozo straight on this matter.

Exercise 9: Resonances again

The chapter makes frequency use of the concept called "resonances." What's the relationship between that concept and resonance as described back in Chap. 3?

♠

Chapter 24
Woodwind Instruments

The family of musical instruments known as the woodwinds includes reed instruments like the clarinet and oboe. It also includes edge-tone instruments like the recorder and flute. Unlike the brass instruments, which are sealed systems of plumbing terminated by a bell where all the sound comes out, the woodwinds have tone holes, and much of the sound is radiated by these tone holes.

24.1 Single-Reed Instruments

A single-reed instrument has a cane reed attached to a mouthpiece. The reed vibrates against the mouthpiece, allowing a puff of air into the rest of the instrument each time the reed opens. If the instrument plays a tone with a fundamental frequency of 440 Hz, the reed opens 440 times per second. Among the well-known single-reed instruments are the clarinets and saxophones.

The Clarinet The clarinet (Fig. 24.1) has a cylindrical bore. Although there is a small bell at the end, the bell is acoustically unimportant except in playing the lowest notes, when most of the tone holes are closed. The clarinet, of all musical instruments, most resembles the cylindrical pipe open at one end and closed at the other. You will recall from Chap. 8 that such a system has only odd-numbered resonances. The low notes of a clarinet reflect that fact. The second and fourth harmonics of tones with low-frequency fundamentals are very weak compared to the third and fifth harmonics. Wow—the physics is working! Contrary to expectation, however, the sixth harmonic can be relatively strong. The wavelength of the sixth harmonic is short enough that small deviations from the cylindrical bore become important—the mouthpiece is tapered, and the bore of the entire instrument is dimpled with closed tone holes.

The low-frequency tones constitute the low "register" of the clarinet. There are 19 notes in this register, and they have the distinctive hollow sound of a complex periodic tone with missing second and fourth harmonics. These notes correspond

W.M. Hartmann, *Principles of Musical Acoustics*, Undergraduate Lecture Notes
in Physics, DOI 10.1007/978-1-4614-6786-1_24,
© Springer Science+Business Media New York 2013

Fig. 24.1 The clarinet resembles a cylindrical pipe, open at one end and closed at the other

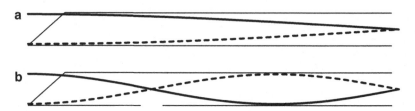

Fig. 24.2 (**a**) Schematic clarinet with the first mode of vibration. (**b**) Schematic clarinet with register hole open supporting the second mode of vibration with a frequency that is three times higher than the first

to the first mode of vibration of air in an open–closed pipe, where the open end is approximately at the position of the first open tone hole.

Higher-frequency tones on the clarinet are in the second register, where the standing waves correspond to the second mode of vibration of air in an open–closed pipe. The second mode has one node, and the clarinet is fitted with a register key, which opens a hole to help force a node in the air column at the right place to activate the second mode of vibration. Because the frequency of the second mode is three times the frequency of the first, the clarinet "overblows the twelfth." The musical interval called the "twelfth" is an "octave" and a "fifth," or a factor of three. (See Appendix D.) Still higher notes on the clarinet use the third mode of vibration of air, a mode with two nodes (Fig. 24.2).

The Saxophone Unlike the clarinet, the saxophone (Fig. 24.3) has a conical bore, i.e., the bore is like a truncated (cut-off) cone. While a cylinder open at one end and closed at the other has only odd-numbered resonances, a cone has both odd- and even-numbered resonances. The resonant frequencies of a cone are given by the formula $f = [v/(2L')]n$, where v is the speed of sound in air, and n is an integer, 1, 2, 3, ... Length L' is somewhat strange. It is not the length of the truncated cone. Instead, it is the length that the cone would have if it were not truncated (Fig. 24.4).

Free Reeds Instruments like the harmonica and accordion also generate musical notes using a single reed. Musical toys played at parties on New Years Eve operate on a similar principle. These instruments might be called woodwinds, but they are not in the same family with the clarinet and saxophone. The clarinet and saxophone have a *beating reed*—the reed opens once per cycle of the tone as it beats against the mouthpiece. The free reed is in a window that allows it to open both in and out of the window. Therefore, the free reed opens twice per cycle of the tone, and the

Fig. 24.3 Saxophones from left to right: soprano, tenor, baritone

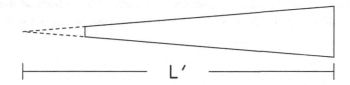

Fig. 24.4 A saxophone is like a truncated cone, shown by the *solid line*. Its effective length is L' including the *dashed conical extension*

spectrum has a strong second harmonic. Another important difference is that the playing frequency of a free-reed instrument is mainly determined by the reed itself. This frequency is established by a combination of stiffness (numerator) and mass (denominator) that you would expect for a free vibrator. By contrast, the playing frequency of a clarinet or saxophone is determined by the air column. The natural frequency of the reed in these instruments does have an effect, but it does not affect the playing frequency, instead it leads to a formant. Harmonics in the tone that are near the reed resonance tend to be strong.

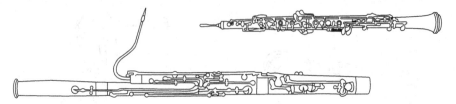

Fig. 24.5 The oboe (*top*) and the bassoon (*bottom*) are the two most important double-reed instruments

24.2 Double Reeds

The best-known double reed instruments are the oboe and the bassoon (Fig. 24.5). In contrast to a single reed instrument, where the reed vibrates against a mouthpiece, the two reeds of a double reed vibrate against one another. At instants in time when there is a gap between the two reeds a puff of air enters the rest of the instrument.

The oboe and bassoon both have conical bores, like the saxophone, but the cone angle is less than in the saxophone and this difference makes the high harmonics of the double reed instruments relatively stronger. Appendix C shows that the frequency range of the oboe is similar to that of the clarinet, though narrower, and the range of the bassoon is similar to the cello. Many of the tones played on these instruments have unusual spectra, with harmonics 3, 4, and 5 as strong or stronger than 1 and 2. Both instruments exhibit formants, near 1,000 and 3,000 Hz for the oboe and near 460 Hz for the bassoon. These formants are probably attributable to reed resonances. As a result, the oboe and bassoon have a very distinctive tone color. Traditionally, the oboe is used to tune the orchestra, possibly because the upper harmonics are strong enough to stand out over the cacophony as other instruments tune up.

24.3 Reeds in General

Although there is a lot of variety among the single and double reed instruments, it's important to recognize that the basic mechanism of tone production for all of these instruments is similar to the process described for a bugle in the chapter on brass instruments. Just as feedback from the horn entrains the vibration of the brass player's lips, so the feedback from the main body of a reed-woodwind instrument entrains the vibrations of the reed. Both lips and reeds operate as pressure-controlled flow valves. The large pressure variations at the mouthpiece (closed end) caused by the standing waves in the body of the instrument control the entry of puffs of air into the instrument. The importance of the body of the instrument becomes evident if the instrument is disassembled, and the reed and mouthpiece are blown in the usual way. The sound of the mouthpiece and reed alone is a high-frequency squeal. The main body of the instrument is needed to support a useful playing frequency. Just as in the brass instruments, the resonances of the plumbing determine the harmonics of the tones, but the player's embouchure (position, orientation, and stiffness of the lips) determines whether the playing frequency will be high, low, or in the middle, and thus determines which resonance will be assigned to each harmonic.

24.4   Edge Tone Instruments

When a stream of air impinges on a sharp edge, the pattern of air flow is unstable. If you blow across the width of a pencil, for example, you will hear a rushing noise coming from the air stream chaotically vibrating from one side of the pencil to another. In musical instruments employing a vibrating air stream as the original source of sound, the vibrations can be entrained by attaching a resonator, such as an open pipe. This is the basis of instruments like the recorder or the flute. The edge, where the tone is generated, is in a hole in the wall of the cylindrical tube. As a result, the instrument resembles a pipe with *two open ends*. Therefore, an edge-tone instrument is different from a reed instrument, where the tone is generated at a *closed end*.

Edges with a Fipple   A whistle, like a referee's whistle, is an edge-tone instrument that anybody can play—even without practicing. That's because it has a fipple, an entry duct that guides the stream of air against the edge. With enough blowing pressure, good tone generation is automatic. Recorders (Fig. 24.6) are instruments with fipples, conical bores, and tone holes. They were popular in baroque times and are still used in baroque music ensembles. Recorders are often used for music education in schools because they can be made inexpensively out of plastic and are easy to play because of the fipple.

The Ocarina   The ocarina is another fippled-instrument that is easy to play. The ocarina is rather unique among musical instruments in that its shape can be just about anything that the maker wants. Typical shapes resemble blobs, and the ocarina is sometimes called the "sweet potato."

Unlike the other instruments of this chapter, which use the modes of air columns, the ocarina uses a mode similar to the mode of a soda bottle. It is called the "Helmholtz mode" or the "Helmholtz resonance." In this mode of vibration, the mass of air in the neck of the bottle serves as the vibrating mass, and the

Fig. 24.6 An alto recorder is an ancient instrument

Fig. 24.7 This Helmholtz
resonator has a clearly
defined neck

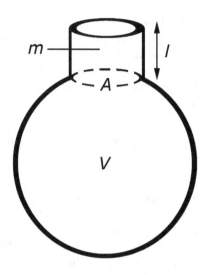

compressibility of air in the rest of the bottle serves as the spring. Of course, for
most bottles the division into mass and spring parts is not clear cut. The bottle tends
to be a continuous system. But just as one could identify mass in the tines of tuning
fork and spring in the shoulders of a tuning fork, one can say that parts of the air in
the bottle play the role of a mass and parts play the role of a spring (Fig. 24.7).

The frequency of the Helmholtz resonance is given by

$$f_H = \frac{v}{2\pi}\sqrt{\frac{A}{V\,l}} \qquad (24.1)$$

where v is the speed of sound in air, V is the volume of the bottle, A is the cross-
sectional area of the neck, and l is the length of the neck. The dependence on the
volume of the bottle indicates that the details of the shape are unimportant.

By blowing very hard, a player can excite a second mode of vibration of air in an
ocarina—more easily in some instruments than in others. Modes of vibration higher
than the Helmholtz resonance are called "cavity modes." In general their frequencies
are much higher than f_H and not at all harmonically related to f_H. The shape of a
cavity mode, its frequency, and the ease with which it can be excited depend on the
shape of the instrument.

A case can be made that the ocarina is a more successful instrument than the
recorder. The problem with the recorder is that it is difficult to play low-frequency
notes loudly. When the player increases the blowing pressure, there is a great risk
that the oscillation of air will jump from the low-frequency mode to a second mode,
with a frequency that is an octave higher. The ocarina suffers less from this problem
because the second mode of vibration is so far removed in frequency and so different
in mode shape. For the same reason the ocarina does not have a rich spectrum. It's
waveform is mostly a sine tone.

24.5   Boatswain's Pipe

From old movies of ancient naval warfare, you will recall the high-pitched sound of the boatswain's pipe. The pipe, or whistle, of the boatswain (pronounced "bosun") had to be audible in a gale everywhere on the ship. It is tuned to about 3,000 Hz, right where the auditory system is most sensitive and well above the spectral range of most wind and waves. The boatswain's pipe uses the resonance of a small, hollow metal sphere with a hole. Because it is metal, the wall of this spherical cavity is thin and one can hardly imagine a neck length. Thoughts like this raise the concept of the Helmholtz resonator without a neck.

The theory of the neckless Helmholtz resonator supposes that the mass of the oscillator corresponds to the air just outside the tone hole. It is essentially the same concept as led to the end correction for the open ends of pipes. The resonance frequency of such a cavity is given by

$$f_H = \frac{v}{2\pi}\sqrt{\frac{1.85a}{V}}, \tag{24.2}$$

where a is the radius of the tone hole.

24.6   The Flute

The flute (Fig. 24.8) is an edge-tone instrument resembling a pipe open at both ends like the recorder. However, the flute does not have a fipple. Instead, the player is responsible for directing the stream of air optimally against the edge to make a tone. Therefore, the flute is more difficult to play than the recorder. Nevertheless, the flute has now replaced the recorder as an important instrument in orchestras and bands. The reason is that the flute player can redirect the air stream so as to avoid the octave jump at high stream velocities. Therefore, the flute player can play low-frequency tones at higher levels of intensity. Actually, loud, low-register tones on the flute tend to have a very strong second harmonic. But even though the second harmonic is strong, the flute player is able to keep the oscillations going in the first mode of vibration so that the fundamental of the tone is still in the spectrum, and the pitch stays in the lower octave.

Because it has a cylindrical bore for most of its length, the flute overblows the octave. "Overblowing the octave" means that if a player is playing a

Fig. 24.8 The flute looks complicated, but it is basically a cylindrical pipe with holes in the wall

low-frequency tone, and blows harder without making a compensating correction in the embouchure, then the frequency will jump up by an octave. Tones in this second register (one node) and tones in the third register (two nodes) are rather like sine tones if played softly. No other instrument in the orchestra makes a tone that is more sinusoidal.

The flute is about 60 cm long. That is such a long piece of pipe that if tone holes are placed optimally, the tone holes are too far apart for a player's fingers to reach. As for the reed woodwinds, an elaborate system of keys, rings, and pads has been worked out so that the instrument can be played with ordinary human hands. The present design was invented by Theobald Boehm in 1871. The world still awaits the genius who will invent such a system for the ocarina. When this happens, the sweet potato will be boss.

The piccolo operates according to the same principles as the flute. It is about half as long, and, as expected, plays about an octave higher.

Exercises

Exercise 1, Clarinet resonances and clarinet tones
The clarinet is approximately a cylindrical tube, open at one end and closed at the other. (a) Show why you expect the clarinet tone to have only odd-numbered harmonics. (b) Appendix C says that the lowest note on a clarinet has a frequency of 147 Hz. Given a first mode frequency of 147 Hz, calculate the expected length of a clarinet.

Exercise 2, Clarinet register key
The clarinet has a register key to aid in playing the second mode of vibration of air in the column. Why would you expect the register key to be at a position that is 1/3 of the way from the mouthpiece to the end of the horn?

Exercise 3, Clarinet and flute
A flute is a pipe open at both ends. It is about as long as a clarinet, but it sounds an octave higher. Why?

Exercise 4, Clarinet and oboe
The oboe has a conical bore. It is not cylindrical like the clarinet but flares on the inside. The oboe also has a double reed and not a single reed. The double reed of the oboe snaps shut for an appreciable fraction of a normal playing cycle. The single reed of the clarinet does not normally close so much except at very high playing levels. Explain why you expect the oboe to have a brighter tone color.

Exercise 5, Reeds and singing
(a) In what way are the behaviors of vocal folds and vocal tracts in singing similar to a free-reed instrument . . . and not similar to a saxophone? (b) In what way are the vocal folds more like an oboe than like a clarinet.

Exercise 6, The flute

From the embouchure hole to the open end, the flute is about 60 cm long. What do you expect the frequency to be for the lowest possible note on the flute?

Exercise 7, Platinum flutes and metal clarinets

The sound of a woodwind instrument comes from the vibrations of a column of air inside the instrument. The walls of the instrument only serve to confine the air column. The walls do not vibrate much, they do not radiate appreciably. Therefore, the scientific evidence supports the notion that the wall material is ... well ... immaterial to the playing of the instrument or the quality of the tone. Musicians on the other hand often disagree. What do you think?

Exercise 8, The bosun's pipe

On a boatswain's pipe, the radius of the tone hole is about half the radius of the spherical cavity. Show that the frequency of the pipe is given by

$$f_H = 5146/D \tag{24.3}$$

where D is the diameter of the sphere in cm. Therefore, to make the pipe have a frequency of 3,000 Hz requires a sphere with a diameter of about 1.7 cm.

Exercise 9, Tuning up to the oboe

The text says that the oboe may be used to tune the orchestra because its upper harmonics are strong enough to be heard against the rest of the orchestra. What good would that be? What good would it be to hear only the upper harmonics?

Exercise 10, Harmonica spectrum

Figure 24.9 shows the power spectrum of the lowest note on a Hohner Blues Harp. (a) What is the fundamental frequency, and what musical note would that be? (See Appendix E.) (b) How many harmonics can you find over the range of the spectrum? (c) Do you expect the harmonica tone to be periodic in time? (d) Compare the levels of the fundamental and the second harmonic. (e) Compare the levels of the fundamental and the 12th harmonic.

♠

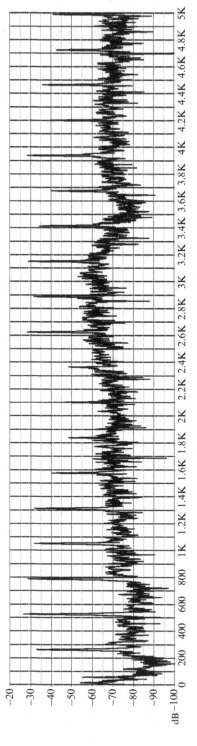

Fig. 24.9 Spectrum of the lowest note on a small harmonica for Exercise 10

Chapter 25
String Instruments

It is hard to overestimate the importance of the stretched string in music—so many instruments, ancient and modern, use the stretched string as the initial source of sound. The string is used in two different ways, as a percussive string and as a bowed string. A percussive string may be plucked, as in a guitar or harpsichord, or it may be struck, as in a piano. A bowed string is continuously excited by the action of the bow and it produces a sustained sound. This chapter treats the percussive and bowed strings in turn.

25.1 Percussive Strings

A percussive string is a free vibrator. After it is once struck or plucked, it vibrates in its natural modes like the vibrators of other percussion instruments such as bells or drums. Chapter 7 described the modes of a freely vibrating string and noted that ideally their frequencies are in a harmonic relationship. Chapter 7 concluded by noting that real strings are not completely ideal and that deviations from ideal behavior would be presented at a later time. That time has now arrived.

Like every oscillator, a stretched string must experience a restoring force that tends to move it back in the direction of equilibrium when it has been displaced. The restoring force comes about in two ways. First, there is the tension in the string, second there is the stiffness of the string. To the extent that the restoring force comes from tension alone, the string is ideal and its modal frequencies are perfect harmonics. To the extent that the restoring force comes from stiffness the string resembles a solid rod. As will be seen in Chap. 26 on percussion instruments, the solid rod is a very inharmonic system. Every real musical string is partly ideal, but partly it is a rod.

The stiffness of a string is not normally important for low-numbered modes of vibration that don't bend the string sharply. However, high-numbered modes of vibration, with many bends, are affected by the stiffness, and their frequencies

W.M. Hartmann, *Principles of Musical Acoustics*, Undergraduate Lecture Notes
in Physics, DOI 10.1007/978-1-4614-6786-1_25,
© Springer Science+Business Media New York 2013

are progressively increased. For the highest modes, the deviation from perfect harmonicity is proportional to the cube of the mode number, i.e., proportional to n^3. A formula that describes this behavior gives the frequency of mode number n as

$$f_n = nf_1[1 + \alpha(n^2 - 1)]. \tag{25.1}$$

The ideal behavior, where f_n is just n times the frequency of the fundamental (f_1), is outside the square brackets. The deviation is within the square brackets.

The size of the deviation from ideal behavior is determined by α, which depends on the properties of the string,

$$\alpha = \frac{\pi^3 r^4 E}{8FL^2}. \tag{25.2}$$

Here r is the radius of the string, and E is a property of the string material. It is a measure of the resistance to being stretched known as Young's modulus. In the denominator, F is the tension in the string, and L is the length of the string. The interpretation of Eq. (25.2), with quantities r, E, L, and F, is that if you want to get ideal behavior then you need a *thin* (r) and *flexible* (E) string. It should be *long* (L) and should be stretched to a *high tension* (F). The opposite conditions tend to lead to a large value of α and more inharmonic behavior.

At this point it is good to reconsider the guitar player's equation, number (7.7), which says that the playing frequency, or fundamental frequency, of a string is

$$f_1 = \frac{1}{2L}\sqrt{\frac{F}{\mu}}, \tag{25.3}$$

where μ is the linear mass density of the string. This equation suggests that if you want to see ideal behavior then it is the low-frequency tones that are going to give you trouble. The reasoning is this: first, there is a practical limit on how long the string can be. (However much you might enjoy a 20-ft piano, you couldn't get it into the house.) Therefore, to get low frequencies you need to reduce the tension (F) and increase the mass per unit length (μ). But reducing the tension and making the string thicker (r) to increase the mass both tend to increase parameter α and lead to inharmonic components in the tone.

One way to help solve the low-frequency problem is to use a wound string. A wound string consists of a thin nylon or steel string core, with windings of brass wire. The windings add mass to the string to help increase μ but they are not nearly as stiff as a solid wire with comparable mass because the windings are in flexible coils. Guitars, pianos, and violins all use wound strings for the low-frequency strings.

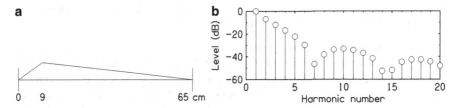

Fig. 25.1 Part (**a**) shows a guitar string, 65 cm long, plucked with a sharp bend 9 cm from the bridge. Because the initial displacement pattern is sharp the pattern includes high-frequency modes of vibration. The initial spectrum in (**b**) includes those high frequencies as harmonics. Because the ratio 9/65 is approximately 1/7, it happens that the 7th and 14th harmonics are weak in the spectrum

25.1.1 The Guitar

The guitar has six strings, tuned to frequencies 82, 110, 147, 196, 247, and 330 Hz. The strings are played by plucking, which means that a string is pulled away from equilibrium with some initial displacement pattern and then released.

The tone color of the guitar depends on the initial displacement pattern. If the guitar is plucked in such a way that the initial displacement is rounded, the tone color will be dull. If the string is plucked with a pick so that the initial displacement has a sharp bend, or kink, the tone color will be bright. It is not hard to figure out why the tone color behaves this way. The initial displacement determines the initial amplitudes of the different modes of the string. Imagine that a string is plucked right in the middle using the broad part of the thumb. This initial displacement looks a lot like the first mode of vibration, with little contribution from high-frequency modes. With little amplitude in high-frequency modes, this sound is dull. If the string is plucked with a fingernail or pick near the bridge, the initial pattern has a sharp kink. To form that pattern by adding up the shapes of the modes of a stretched string requires modes with short wavelengths. Therefore, high frequencies will be included in the tone and the tone will be bright, as indicated in Fig. 25.1.

The waveshape and spectrum in Fig. 25.1 look like an unsymmetrical triangle wave, the kind of wave that could be produced by a function generator. But the guitar does not sound at all like a function generator. The difference lies in what happens after the initial sound. As the guitar tone decays away it evolves in several ways.

1. The harmonics of the tone do not decay at equal rates. The higher-frequency modes are damped more than the lower-frequency modes causing the high harmonics to decay more rapidly than the low. The waveform on the string becomes smoother as the decay progresses.

2. The harmonics are not truly harmonic. Because the plucked string is a free vibrator, stiffness in the string causes the high-frequency components to be slightly higher in frequency than true harmonics of the fundamental. Because the frequency shift is not large, we still speak of harmonics, but the shift causes the relative phases

among the harmonics to change with time, and this change is another reason that the shape of the wave evolves as the tone sounds.

3. There may be audible effects because the vertical and horizontal polarizations of the string vibration do not have identical frequencies. Recall that the string displacements are transverse to the string length, but these may be either vertical (perpendicular to the guitar top) or horizontal (parallel to the guitar top). Because of the connection of the string at the bridge, the string appears to have slightly different lengths for these two polarizations and the difference may produce slow beats, a further kind of evolution.

4. Most important, the string vibration is not heard directly. The string itself moves very little air. Instead, in an acoustic guitar, the string is coupled to the top plate through the bridge, and the top plate is coupled to the air cavity within the guitar. The top plate radiates because it has a large surface, and the cavity radiates through the sound hole. The top plate and cavity together have resonances that filter the sound of the string as they radiate it, in much the same was as the vocal tract filters the glottal pulse from the vocal folds.

25.1.2 The Electric Guitar

The electric guitar is different from the acoustic guitar in that the sound is not radiated by the top plate and sound hole. In fact, a top plate and sound hole don't normally exist, and if they do they do not matter acoustically. Instead, the string vibrations are picked up by electromagnetic transducers to create an electrical signal which can then be amplified (often a lot) and reproduced by loudspeakers (often very large loudspeakers).

A transducer is made by wrapping coils of wire onto a permanent magnet beneath a string. The string is steel, which is a magnetic material. Thus when the string vibrates it alters the field from the permanent magnet and creates a voltage in the coil by means of the generator principle described in Chap. 16. The effect generates an appreciable voltage only if the coil is close to the vibrating string, and there is normally a separate transducer for each string.

The signal that comes out of the electric guitar represents the string motion at the position of the transducer. If the transducer is placed near the bridge, it efficiently picks up high-frequency modes and the tone is bright. If the transducer is farther from the bridge it picks up the large vibrations of low-frequency modes and the tone is mellow. Electric guitars often have two or three sets of transducers at different distances from the bridge to provide options.

Except for this very different sound radiation process, the electric guitar is like an acoustic guitar. Similar processes cause the string vibration to evolve in similar ways as the tone decays, although the decay is of longer duration in the electric guitar (see Exercise 5).

25.1.3 The Piano

The piano is like the guitar in that the strings of a piano are free vibrators. However, while the guitar strings are plucked, the piano strings are struck. A modern piano has 88 keys. When a key is depressed, a hammer with a felt surface is thrown against the strings tuned to a single note. The striking of the strings leads to a brief noise at the very start of the piano attack. As in the guitar, the strings themselves radiate very little sound. The strings of the piano pass over a bridge connected to a soundboard, and the soundboard is responsible for the radiation.

As for the guitar, the frequencies of the modes of vibration of the string are "stretched" compared to harmonic frequencies. There are similar questions of scaling. The string length, string tension, and string mass must be chosen to play the right frequency, to obtain good coupling to the radiator, and to satisfy practical considerations such as the size of the instrument and the physical limits on the maximum tension. The piano maker has some advantages. The musician does not hold the piano in his lap, and the piano can be large. The frame can be made of heavy steel to support the tension in the strings—a total of 20 tons of tension in a concert grant piano! Of course, there are correspondingly great demands on the piano maker. The piano needs to cover a wide frequency range—notes tuned from 28 to 4,196 Hz, and the piano is required to produce a big sound.

The three effects of a non-ideal string that lead to an evolving vibration pattern, and an interesting tone, for the guitar also apply to the piano. In addition, the piano tone has another avenue of evolution. Piano tones are played on multiple strings. Corresponding to the 88 different notes on the piano are more than 225 strings. Most of the notes of the piano have three strings. The mid-bass region has two strings, and only the low bass notes have a single string.

The three strings assigned to a single piano tone are not all tuned to identical frequencies. First, it would be physically impossible to make them absolutely identical. Second, there is evidence that piano tuners adjust the mistuning among the three strings to produce the best sound. Mistuning leads to beats, of course, but the different harmonics of the tone beat at different rates and the perceived effect of the beats is subtle. The mistuning among the multiple strings leads to an important characteristic of piano tone, a double decay pattern.

When the hammer strikes the three strings, it first causes them all to move up and down together. All the strings force the bridge in the same way. The bridge moves a lot and extracts energy from the vibrating strings with great efficiency. This leads to an intense attack sound. It is a sound that decays rapidly because the strings are losing energy at such a high rate. But, because the strings have slightly different frequencies, it takes only a few seconds before the vibrations of the individual strings are no longer in phase. Then the strings force the bridge independently. This causes the coupling to the bridge to be less efficient, and the rate of decay becomes slower. The slowly decaying part of the tone is called the "aftersound." It can easily continue for a minute (Fig. 25.2).

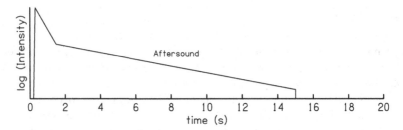

Fig. 25.2 The piano tone has a two-part decay pattern. At first, the strings vibrate in phase, creating an intense rapidly-decaying tone. After the string vibrations become out of phase, the aftersound decays more slowly. The tone shown here stops abruptly after 15 s because the player released the key. When the key is released a damper makes contact with the string and stops the tone

Fig. 25.3 Bowed string instruments, from *left* to *right*: the violin, the viola, the cello, the doublebass or bass viol

25.2 Bowed Strings

The class of orchestra instruments known as the strings includes the violin, viola, cello, and double bass (Fig. 25.3). These instruments are sometimes called the violin family. The ranges of these important instruments are given in Appendix C. The basic sound producing element is the vibrating string. Sometimes the string is plucked, a style of playing known as "pizzicato." When plucked, a violin string is a free vibrator, just like a guitar string. Normally, however, the string is bowed.

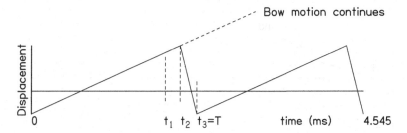

Fig. 25.4 Two cycles of the stick–slip process. The y axis shows the displacement of the string at the location of the bow as time progresses. The slipping takes place between times t_2 and t_3. Because it takes 4.545 ms for two cycles, you know that the playing frequency is 440 Hz. Time t_1 is a marker for Fig. 25.5

Bowed strings are not free vibrators. They are driven vibrating systems like the brasses and woodwinds. Like the brasses and woodwinds, the bowed strings are sustained-tone instruments. Their tones have exact harmonics and are stable when viewed on an oscilloscope. As in all the other sustained-tone instruments, feedback and a nonlinear tone generation process are responsible for creating a tone with harmonic partials from resonances that are only approximately in a harmonic relationship.

25.2.1 Tone Generation in the Bowed Strings

The tone generation process is similar in all the bowed string instruments. For definiteness, the following discussion will refer to the violin, but it might as well be directed toward any of the others.

The bow of a violin is made of many strands of horse hair, stretched into a thin ribbon. When the bow is drawn across a string, the string is set into vibration. The process by which the bow excites the string can be described as a "stick–slip" process. As the bow moves across a string, it pulls the string with it. This is the "sticking" phase of the stick–slip process. Violin players put rosin on the bow to control the amount of sticking. As the string is pulled aside by the bow it is stretched, and tension builds up. Eventually the tension is so great that the string slips and moves in a direction that is opposite to the bow motion. This is the "slipping" phase of the stick–slip process. Once the string has slipped back, it is ready to be picked up by the bow again and another cycle begins. During the sticking part of the cycle, the speed of the string is the speed of the bow. But when the string slips, it snaps back suddenly, as shown in Fig. 25.4.

This stick–slip description of the bowing process is easy to understand. What is more, it correctly describes the motion of a bowed string as seen at the bowing point. However, there is something dreadfully incomplete about it. It takes no account of the length of the string. But you know that the length of the string is crucial in

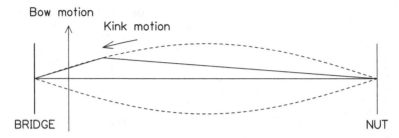

Fig. 25.5 When a string is correctly bowed, the string displacement has a sharp kink which travels around the path shown by *dashed lines*, once per cycle. A snapshot of the string is here taken at time t_1 (see Fig. 25.4), just before the kink hits the bowing point. When the kink hits the bowing point the slipping phase will start

determining the playing frequency. A violin player shortens the string by pressing it down on the fingerboard to make tones with higher frequency. So there is a puzzle here to solve.

Perhaps you are now thinking about how it was with woodwind instruments— how resonant feedback from the body of a woodwind instrument entrained the opening and closing of the reed, which then fed energy into the oscillating system. That feedback was critical in determining the playing frequency of the instrument. Perhaps you are now thinking by analogy and imagining that feedback from the well-known resonances of a stretched string, might entrain the stick–slip process. If that is what you are thinking, you would be right. It is absolutely uncanny how this general physical principle keeps on appearing, in different guises, in the different sustained-tone instruments.

If you look down on a string while it is being bowed, you will see a fuzzy pattern looking like the dashed outline in Fig. 25.5. This outline is actually the path of a sharp bend, or kink, in the string, as shown by the solid line. This kink travels around the path once per cycle. If the playing frequency is 440 Hz, the kink makes 440 round trips per second. Figure 25.5, is a snapshot of the string taken at time t_1, just before the kink arrives at the bowing point. An instant later, at time t_2, the kink hits the bowing point and knocks the string free from the bow. That initiates the slipping phase, as shown in Fig. 25.4. While the string is slipping, the kink will travel to the bridge, where it will be reflected onto the bottom dashed path in Fig. 25.5. When the kink hits the bowing point again, the bow will pick up the string and the sticking process will begin again. In that way, the modes of the string make their impact on the stick–slip mechanism.

The stick–slip process is periodic, and the period is equal to the time it takes for the kink to make a round trip. In one cycle the kink travels a total distance of twice the string length ($2L$), and it travels with the speed of sound on the string (v). Consequently, the period is $2L/v$, and the fundamental frequency is $v/(2L)$. It should not surprise you that this frequency is exactly what we found in Chap. 7 for the fundamental frequency of a stretched string.

Fig. 25.6 A violin bridge
with the pitch-names of the
strings indicated

This account of the vibrations of a bowed string describes what happens when
the string is correctly bowed and a good tone is produced. It is worth noting that
there is no guarantee that just any application of a bow to a string will be correct or
that a good tone will be produced automatically. Getting a string to vibrate so that
there is useful cooperation between the energy fed in from the bow and the feedback
from the string itself is the art of bowing a string, and every violin player learns to
do that. In fact, if a string is bowed with too much bow force, the feedback process
fails. Then the stick–slip process is uncontrolled, and the string produces a noisy
squawk with no clear frequency.

25.2.2 The Violin Body

In a wind instrument, the air column plays a double role. It provides a resonant
feedback mechanism responsible for playing a periodic tone with a stable frequency,
and it is responsible for radiating the tone, either through a bell or through tone
holes. The violin is not like that. As described above, the resonant feedback
mechanism is the stretched string and does not involve the body, but a string makes
a very inefficient radiator. It is so thin that it cannot move much air. What happens
instead is that the violin string causes the bridge to vibrate, and the bridge causes
the top plate of the body to vibrate in turn. The coupling of the vibration through
the bridge is a very delicate operation. The bridge is carved in a special shape that
allows it to provide a good match between the string and the top plate. The *mute*
on a violin, for especially soft playing, adds a little mass to the bridge and greatly
reduces the coupling efficiency for high-frequency vibrations. The fact that a small
change in mass in the bridge dramatically reduces the transfer of high-frequency
power indicates that the function of the bridge is highly sensitive to its physical
details (Fig. 25.6).

The radiation from the violin comes mainly from the top plate and from the
vibration of air inside the body. Holes, called "f-holes," are cut into the top plate and
allow sound waves inside the body to radiate to the outside. The resonances of the
violin body and air volume are independent of the modes of the strings, and they are
of the utmost importance for the sound quality of a violin. They mark the difference
between a poor sounding student instrument and a great sounding masterpiece.

It is thought that the art of violin making reached its peak, in Italy, at the time of Niccolo Amati (1596–1684) and Antonio Stadivari (1644–1737). In the centuries since then countless violin makers (and scientists too) have tried to make instruments of comparable quality. It has been supposed that these Italian craftsmen held a secret—perhaps it was a special varnish—and that one only had to discover the secret to achieve the playing characteristics and sound of these splendid instruments. Possibly all this is true, and yet there is more than just a little mystique about these instruments. Most of the fine old Italian violins that are so prized have been rather thoroughly worked over in more recent times to keep them in playing condition. One wonders how much of the original is left.

Whether or not contemporary violin makers can make violins that compare with the old masters can be debated. What can be said for sure is that contemporary makers know enough about the art and science of violin making to make fine playing instruments and can do so reliably.

Vibrato   A violinist does not press the string down on the fingerboard and hold it there motionless. Instead, the violin player oscillates the hand (it's always the left hand) so as to make the string slightly shorter and longer with time. This leads to a playing frequency that rises and falls periodically. The effect is well characterized as frequency modulation (FM) of the violin tone. The parameters of this FM are different for different playing styles and are subject to fashions of the times. In renaissance music there was no vibrato at all. Over the course of the twentieth century preferred vibrato rates decreased from a range of 6–8 cycles per second to a range of 5–7 cycles per second. For a contemporary violinist typical parameters are: a rate of 5.5 cycles per second, a sinusoidal modulation waveform, and an overall frequency excursion of 1 semitone (equivalent to $+3\%$ and -3% of the playing frequency).

Vibrato is said to be an "ornament" in performance. That statement is not false, but it may not do justice to the importance of vibrato in the sound of the violin. It has become an essential element of playing, so much so that when a violinist has the possibility of playing a sustained note on an open string (no finger down on the fingerboard) the violinist will choose to play the note on a different string where vibrato can be used.

A violin tone with vibrato is lively. Its character is more interesting than a violin tone played without vibrato. The difference may well be that vibrato leads to an effect known as "FM-induced AM." In this effect, the harmonics of the violin tone rise and fall with the frequency excursions in the vibrato. For example, imagine a violin body with a strong resonance at 800 Hz. Imagine now a violin tone with a nominal frequency of 392 Hz, but excursions that make the instantaneous fundamental frequency vary from 380 to 404 Hz. The second harmonic will vary in step with the fundamental from 760 to 808 Hz because the waveform remains periodic throughout the vibrato cycle. When the second harmonic is near 800 Hz its amplitude will be boosted by the resonance, but when the second harmonic is near 760 Hz its amplitude will be weaker. While this second harmonic has *increased* in amplitude with increasing playing frequency, there will no doubt be other harmonics

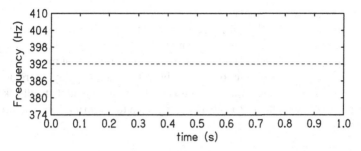

Fig. 25.7 The note being played is a *G* with a central frequency of 392.0 Hz. One second of the sound is shown here. The *vertical axis* is intended to show the instantaneous frequency to be drawn in Exercise 4

of the tone that line up with different resonances of the body in such a way that their amplitudes *decrease* with increasing playing frequency. Because the violin body has many resonances, there are many opportunities for this effect. The dynamic character of FM induced AM leads to a shimmering quality to the violin tone.

Exercises

Exercise 1, Speed of sound on a violin string
     A violin string is 346 mm long, from bridge to nut. If the string is tuned to 440 Hz, what is the speed of sound on the string?

Exercise 2, How to play the violin
     An open violin string is 346 mm long and plays 440 Hz. To play higher notes the violinist shortens the string. Each semitone step of the ascending scale corresponds to an increase of about 6 % in frequency. Thus, successive notes of the chromatic scale have frequencies of about 440, 466.2, 493.9, 523.3, 554.4, 587.3, 622.3, and 659.3 Hz, as given in Appendix E. Calculate the string lengths for all of those tones. How close are these string lengths to being equally spaced on the fingerboard?

Exercise 3, The violin and the voice
     The violin sound begins with a string and is radiated by the body. The human voice begins with vocal folds and is radiated by the vocal tract. Find the parallels between the violin and the voice.

Exercise 4, Vibrato
     On Fig. 25.7 sketch the instantaneous frequency of a violin tone played with vibrato that is acceptable by contemporary standards.

Exercise 5, Guitars
     The section on electric guitars says that the string vibration decays more slowly in an electric guitar than in an acoustic guitar. Why is this so?

Exercise 6, Bozo's string theory

Bozo says that the stick–slip process for bowing a string predicts an effect of adding rosin to the bow. He says that adding rosin makes the bow stickier so that the string continues to be dragged by the bow for a longer time. That makes the period longer. The result is that adding rosin to the bow decreases the playing frequency.

Use what you know about the reflected kink on the string and its feedback to the bowing process to explain to Bozo why adding rosin to the string does *not* affect the playing frequency.

♠

Chapter 26
Percussion Instruments

This chapter is about percussion instruments. The percussion family includes well-known instruments like drums, cymbals, and bells—and also hollow logs, tin cans, and truck springs. Anything that can be beaten or rattled to make a noise is potentially a percussion instrument. Therefore, this could be a very long chapter. To bring a little order to the zoo, this chapter starts with a definition.

Percussion Instruments Defined Percussion instruments are free vibrators. Unlike sustained tone instruments, where power is continually fed into the instrument to keep the tone sounding, a percussion instrument gets all its energy with an initial strike. After the initial impulse the percussion instrument vibrates in its natural modes of vibration. Eventually friction and loss of energy to radiation damp these vibrations so that they can no longer be heard.

It is a fact of mechanics that no real physical system has modes of vibration with frequencies in a precise harmonic relationship. Therefore, a corollary of this definition is that all percussive sounds are more or less inharmonic, or aperiodic. That is an absolutely firm conclusion. On the other hand, the phrase "more or less" gives a lot of latitude. Some percussive systems, such as the percussive strings of guitars and pianos, have modes of vibration that are nearly harmonic. Such systems can play melodies and close harmony just like sustained-tone instruments. Other percussion instruments make sounds that are grossly inharmonic.

Some percussion instruments, such as marimba bars and chime tubes, start with a very inharmonic system but incorporate refinements intended to emphasize their tonal character and enable them to play melodies and even close harmony. Other percussion instruments, like snare drums, go the other way and incorporate features designed to make them less tonal.

To bring further order to the zoo, this chapter continues by organizing percussive instruments according to type of vibrator. It deals with bars, membranes and plates.

W.M. Hartmann, *Principles of Musical Acoustics*, Undergraduate Lecture Notes in Physics, DOI 10.1007/978-1-4614-6786-1_26,
© Springer Science+Business Media New York 2013

26.1 Bars, Rods, and Tubes

A bar is a length of solid metal or wood with a rectangular cross section. Rods (solid) and tubes (hollow) have circular cross sections. Musically useful vibrations of bars, rods, and tubes have vibrations that are transverse to the length. The ends may be free or clamped. For instance, a marimba bar has both ends free, but a clock chime rod has one end clamped.

The modes of vibration of a free bar, rod, or tube can be described by the number of nodes. The first mode has two nodes, the second mode has three, and so on, as shown for a bar in Fig. 26.1.

The frequencies of these modes of vibration are given by the formula

$$f = 0.113 \frac{v_L t}{L^2} (2n + 1)^2, \tag{26.1}$$

where v_L is the speed of sound in the bar material, t is the thickness of the bar, and L is its length. The quantity n numbers the modes, and $(2n + 1)$ takes on the values 3.011, 5, 7, 9, ... for mode numbers 1, 2, 3, 4, ... Thus, it is the sequence of odd integers starting with 3, except that the first value is a little more than 3. It should be evident that the frequencies are not at all in a harmonic relationship. Exercise 1 shows that the second mode of vibration has a frequency that is 2.76 times the first, i.e., $f_2/f_1 = 2.76$. The third mode has a frequency that is 5.40

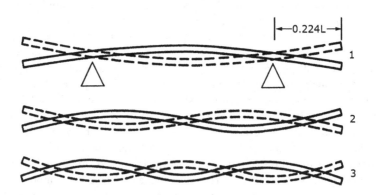

Fig. 26.1 The first three modes of vibration of a uniform rectangular bar with free ends. *Triangles* show mounting points that encourage the first mode and discourage all the others by damping their vibrations

Fig. 26.2 How a bar can be undercut to lower the frequency of the first mode

times the first. It is typical of systems that have stiffness (like bars and rods) for the mode frequencies to grow more rapidly than the mode number. Here the frequencies grow as the square of the mode number. One says that compared to a harmonic relationship, the mode frequencies are "stretched." In musical applications, the pitch that is assigned to a bar corresponds to the frequency of the first mode.

26.1.1 Useful Bars

The marimba, xylophone, and vibraphone are instruments that use rectangular bars to make a tone. These instruments are intended to play melodies and close harmony, and so it is evident that something must be done about all that inharmonicity. In fact, three things are done. First, the bar is undercut, as shown in Fig. 26.2. This lowers the frequencies of all the modes, but especially of the first mode, where all the bending takes place near the middle of the bar. By lowering the frequency of the first mode, the ratio f_2/f_1 can be made 3.0 or 4.0 instead of 2.76. This allows the second mode to create the third or fourth harmonic of the tone. Second, the bar can be mounted so as to favor the first mode of vibration. Figure 26.1 shows that the nodes on the extreme left or right of the bar move toward the outside of the bar for higher mode numbers. If the bar is mounted where the first mode has a node and is not vibrating anyway, the first mode is not damped by the mounting as much as the higher modes. Finally, resonating tubes (pipes open at one end and closed at the other) can be placed under the bars to emphasize the first mode frequency. In the end, the tone produced by these instruments is rather like a decaying sine tone because of the emphasis given to the first mode, but the excitation of all the other modes at the onset makes the sound interesting.

26.1.2 Useful Tubes

Tubes are used to make chimes (Fig. 26.3). Orchestral chimes are a set of long brass tubes that hang in a rack at the back of the orchestra in the percussion section. Wind chimes hang from trees in the back yard to entertain(?) the neighbors. Because these tubes are cylindrical and hollow, you might imagine that the physics of open pipes from Chap. 8 would be relevant here, as they are for the vibraphone resonators. Do not be fooled. The vibration of a chime bar has nothing at all to do with the vibration of air in a pipe. In a chime, the vibration is in the walls of the tube itself, and the relevant speed of sound is the speed of propagation of a displacement of the brass walls. This is very different from a wind instrument or resonant pipe, where the vibration is in the air column and the relevant speed of sound is the speed of sound in air.

The modes of a chime are similar to the modes of a solid bar or rod (see Fig. 26.1) and the frequency ratios of the modes are similar too (see Eq. (26.1)).

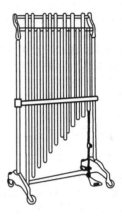

Fig. 26.3 (*Left*) A vibraphone has undercut bars and resonating tubes under the bars. Disks within the tubes are rotated by a motor to tune and detune the resonators. This imparts a tremolo to the sound, favored in jazz applications. Chimes, on the *right*, also have metal tubes but they are used in an entirely different way

The suspension of the chime tube, and the reduced internal friction, allows higher frequency modes to be sustained longer in the chime compared to the solid bar.

What is interesting about the chime is that its nominal frequency is not even close to any of the mode frequencies. Instead, the pitch of the chime comes from modes 4, 5, and 6. According to Eq. (26.1) the frequencies are proportional to 9^2, 11^2 and 13^2, or 81, 121, and 169. These numbers are in the ratios 2, 2.99, and 4.17. These are close enough to the ratio 2, 3, 4 that the chime creates a tone with a missing fundamental, as described in Chap. 13 on pitch perception. The frequencies of modes 4, 5, and 6 can be brought into an even better 2, 3, 4 ratio by loading an end of the chime tube with a solid plug, and manufacturers regularly do this.

26.2 Membranes

Drum heads are membranes stretched over a hollow frame. A few examples appear in Fig. 26.4. They are always circular, or nearly. The modes of vibration of a drum head are displacements perpendicular to the membrane surface, as a function of position on that surface. Therefore, the modes are two dimensional. That makes them different from the modes that have been considered for a stretched string, or for air in a pipe, or for a solid bar, all of which are one dimensional. For a one-dimensional systems there is a mode number, always giving some indication of the number of nodes in the standing wave pattern. For instance, index n describes the number of nodes in Fig. 26.1. To describe a mode in two dimensions requires *two* mode numbers.

Fig. 26.4 A few drums: from left to right: Timpani or kettledrum, Snare drum, Bongo drums (played with the fingers), Bass drum

It is useful to study the membrane that has the greatest possible symmetry, a membrane that is a perfect circle, has uniform density, and is stretched with uniform tension. A well-tuned timpani (kettledrum) with a mylar head approximates such an ideal system. The modes are described by a pair of numbers $[m, n]$, where m gives the number of nodal lines and n gives the number of nodal circles, including the outer rim.

26.2.1   Chladni Patterns

You can see the individual modes of vibration of a drum head by forming Chladni patterns. The first step is to put the membrane in a horizontal orientation and drive it with a steady sine tone from a loudspeaker underneath. Because it has only a single frequency, the sine tone will excite no more than one mode. By tuning the sine tone you can excite the different modes in turn. The next step is to sprinkle aluminum filings onto the membrane. (You could also use glitter, or any dark granular material that is not magnetic and does not clump.) The filings make it easy to know when the sine tone frequency has hit one of the modal frequencies because the membrane will vibrate a lot, and the filings will dance around on top of it. After a short time the filings will settle on the nodal lines and circles. That makes it easy to identify the mode, and a frequency counter attached to the sine tone generator indicates the frequency of the mode.

If you get the chance to do this experiment, it is likely that your drum head will not have absolutely uniform tension. Then the patterns you find will be distortions of the patterns in Fig. 26.5. However, they should be identifiable. Even though a nodal line may curved and a nodal circle may turn into an oval, or acquire cusps, you can tell what the pattern is trying to be.

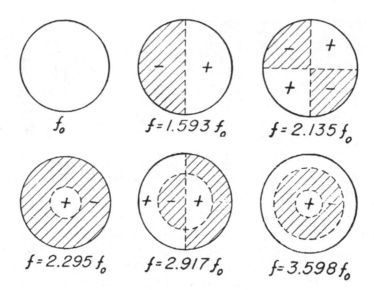

Fig. 26.5 The modes of a uniformly stretched circular membrane are given by two numbers indicating the number of nodal lines and the number of nodal circles. Numbers below the mode drawings show mode frequencies relative to the frequency of the lowest mode, called "[0,1]." Other modes shown are, in order, [1,1], [2,1], [0,2], [1,2], [0,3]

26.2.2 Timpani

Timpani are the most important drums in the orchestra. They are tuned. Although they do not normally play melodies, they are used to reinforce selected notes played by the rest of the orchestra, and it is important that the tuning be right. Here we have another case where a tuned percussive sound is needed and yet the basic vibrator, in this case the stretched membrane, has modal frequencies (as given by the ratios in Fig. 26.5) that do not look promising. The solution to this problem for the timpani is to add the kettle below the membrane. The kettle is filled with ordinary air, and this volume of air loads the membrane and reduces all the modal frequencies. However, some modes have their frequencies more reduced than others and that's the key.

The air loading changes the frequency of the [2,1] mode so that the ratio f_{21}/f_{11} changes from 1.34 to 1.51, and it changes the frequency of the [3,1] mode so that the ratio f_{31}/f_{11} changes from 1.67 to 1.99. The ratios 1.51 and 1.99 are close to 1.5 and 2.0, which are related to the fundamental pitch f_{11} by the interval of a musical fifth and an octave. In this way the kettle of air makes the timpani sound more tonal.

Fig. 26.6 Cymbals (**a**) and gongs (**b**) begin with a free circular plate, but are formed with a dome in the center

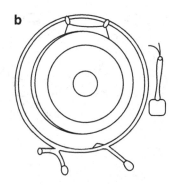

26.3 Plates: Cymbals, Gongs, and Bells

The vibrational modes of a flat, uniform, circular plate are like the modes of a circular membrane in that they can be defined by nodal lines and nodal circles. The same notation, [m,n], can be used. There is a useful analogy that can be drawn between these two-dimensional systems and one-dimensional systems, namely that a plate is to a stretched membrane as a bar is to a stretched string. While the membrane and string have fixed boundaries, the plate and bar (as in cymbal and xylophone) have free boundaries. While the membrane and string get most of their restoring force from the applied tension, the plate and the bar get their restoring force from stiffness. Stiff vibrators have modes with frequencies that are widely separated. This is true of both bars and plates. For comparison, note that the lowest six modal frequencies for a membrane, given in Fig. 26.5, have ratios 1, 1.59, 2.14, 2.30, 2.65, and 2.92. The lowest six modal frequencies for a circular plate with a free boundary have ratios 1, 1.73, 2.33, 3.91, 4.11, and 6.30 (Fig. 26.6).

The circular plate with a free boundary is a reasonable starting model for a cymbal or gong, but both instruments differ in important ways from this prototype. Both cymbal and gong have a raised dome in the center which means that the modes of vibration higher than [0,6] or [2,1] are not like those of a plate but are like combinations of different plate modes. The gong has a curved rim that can be expected to change the free boundary condition. The instrument known as a "tamtam" is a large gong without the dome.

26.3.1 Nonlinear Mode Coupling

The description of the vibration of cymbals, gongs, and tamtams in terms of their natural modes of vibration is a linear analysis and does not really do justice to the complicated character of either the vibration or the sound made by these instruments. The modes are coupled in ways that depend on how hard the plate is struck. For a large gong or tamtam, the vibration pattern develops slowly with time

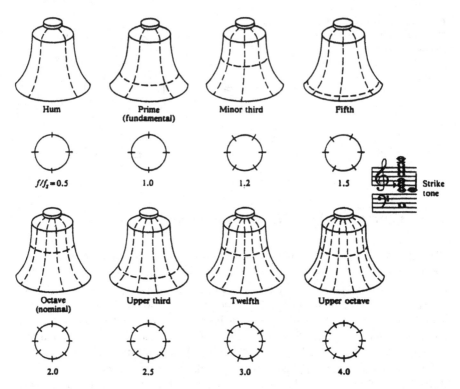

Fig. 26.7 The modes of vibration of a bell have been given names. Their nodal meridians and nodal circles are shown by *dashed lines*. Their frequencies are given relative to the prime frequency which corresponds to the nominal pitch of the bell. This figure was borrowed from *The Science of Sound* by Thomas D. Rossing, Addison Wesley, 1990

as the plate is repeatedly struck. Beating softly on a tamtam with a soft beater leads to a low-frequency rumble. Continued and more intense beating appears to cause a transfer of energy from low-frequency modes into high-frequency modes producing a loud shimmering sound. Psychologically, the effect is dramatic. Intellectually, the scientist also finds this unexpected mode coupling to be dramatic because it is not readily understandable. Whoever figures out how this effect works is going to make scientific history—big time.

26.3.2 Bells

Many civilizations, in the east and in the west, have used bells throughout history. The church bell, or carillon bell, that we know in the west was developed in the low countries of Europe in the seventeenth century.

Figure 26.7 shows that the modes of vibration have nodal meridians and nodal circles that recall the nodal lines and nodal circles of a circular plate. From left to right, the top line of the figure shows modes [2,0], [2,1], [3,1], and [3,1a]. What is different about the bell is that the thickness of the walls is not uniform. Instead, the thickness is caused to vary so that the modes are tuned, with relative frequencies corresponding to notes of the scale, such as the minor third and fifth. Of course, when the bell is struck, all the modes are excited at once and a listener does not hear the individual notes of the scale. The pitch of the bell, or fundamental, is established by the modes with frequencies 2.0, 3.0, and 4.0 times the fundamental. The prime tone acts to reinforce this pitch, and the "hum tone" is consonant, an octave lower. But the modes of the bell decay at different rates, and eventually a listener can hear several different notes of the scale. That ambiguity is part of the charm of the bell sound.

Exercises

Exercise 1, Modes of a uniform bar
Use Eq. (26.1) to find the frequencies of the first four modes of vibration of a uniform bar. Express these frequencies as multiples of the frequency of the first mode. In other words, find f_2/f_1, f_3/f_1, and f_4/f_1.

Exercise 2, The chime tone
It is said that the modes of a chime tube are proportional to 9^2, 11^2 and 13^2, which are in the ratio of 2, 2.99, and 4.17. Show that this is true.

Exercise 3, Drums with two heads
Bass drums and snare drums often have two heads. One is beaten and the other is forced to vibrate by the compression of air inside the drum. The two heads are normally stretched to different tensions. Do you think that the addition of the second head makes the drum more tonal or less tonal?

Exercise 4, Plates you have known
Some dinner plates are made of china, others are made of paper. They sound very different when struck. What is responsible for the difference?

Exercise 5, Bell mode notation
What are the values of $[m, n]$ for the bottom row of Fig. 26.7?

Exercise 6, Bell modes
On opposite sides of a nodal line the vibration is in opposite directions. For instance the nodal meridians in Fig. 26.7 divide the bell into segments. When one segment moves outward, the adjacent segment moves inward. Show how this applies to the circles in Fig. 26.7 that indicate the nodes on the rim of the bell. Use solid and dashed lines—as usual for standing waves—to indicate vibrations as separated in time by half a period.

Chapter 27
Electronic Music

In the broadest terms, the title "Electronic Music" could mean many different things. It could refer to electronic recording of music or to electrified musical instruments like the electric guitar. It could include computer-aided music composition or electronic production of a hard copy of a score. As commonly understood, however, electronic music refers to the electronic synthesis of performed music.

The idea of electronic music stems naturally from the idea of recorded music, either analog or digital. Imagine that you have in your hands a tape recording of a musical performance. You know that the music is entirely captured in the sequence of magnetization stretched out along the tape. It might then occur to you that if there were some way to duplicate that sequence of magnetization, starting from scratch, the original musical performance would not be necessary. Alternatively, imagine that you are holding a compact disc. You know that the music is entirely represented by a long string of ones and zeros on the disc. If you had a computer program that could create the same sequence of ones and zeros, you would have the music, even though it might never have previously existed in any acoustical form. These are the ideas of electronic music—analog and digital.

27.1 Analog Synthesizers

As you know, the first step in recording music is to use a microphone to convert an acoustical signal into an electrical signal. The electrical signal can be perfectly represented by a graph showing the electrical voltage as a function of time. The variations of voltage with time are analogous to the original variations of pressure with time. Because of this analogous relationship, the electrical signal has evident musical value. The goal of the analog synthesizer is to create an electrical signal with musical value without having to record it. Instead, the musician creates the electrical signal from a collection of electronic signal generating and processing circuits.

W.M. Hartmann, *Principles of Musical Acoustics*, Undergraduate Lecture Notes in Physics, DOI 10.1007/978-1-4614-6786-1_27,
© Springer Science+Business Media New York 2013

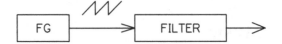

Fig. 27.1 An attempt to synthesize a violin tone using an analog patch of a function generator making a sawtooth wave and a filter with resonances

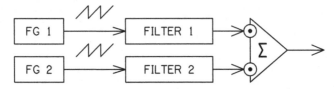

Fig. 27.2 Two tones can be added in a mixer (indicated by Σ to simulate a duet of instruments. The mixer adds the two voltages—more or less depending on the settings of the volume controls, shown by *circular knobs*

Chapter 4 on instrumentation introduced the function generator. The function generator creates a periodic electrical signal with a frequency that can be controlled by a knob on the front panel. For instance, the function generator can create a sawtooth wave. In fact, the sawtooth wave would not seem to be so very different from the wave of a violin sound. (See Fig. 25.4.) Both waves are periodic and both have a complete set of strong harmonics. If you put the fundamental frequency in the right range, perhaps you could use the sawtooth waveform to simulate a violin. However, you will recall that the body of the violin has resonances that emphasize harmonics having particular frequencies. This effect contributes in an important way to the violin sound. Unfortunately for our electronic musical experiment, the sawtooth wave does not have such emphasized harmonics. But there is no need to be discouraged, because it is possible to filter the sawtooth waveform with an electronic filter having resonances. Then the electronic filter changes the spectrum of the sawtooth to simulate the way that the violin body filters the spectrum of vibrations from the bridge. Thus, a first attempt at synthesizing a violin looks like the patch in Fig. 27.1. A *patch* is just a pattern of connections among modules.

The flexibility of analog electronics becomes evident if you want to generate the sound of two violins playing together. You only need to get another function generator and another filter and add the signals in a mixer. The mixer will add the two tones in a linear way, just like the air adds the tones from two violins. To control the levels of the two synthesized tones, there are volume (gain) controls on the inputs to the mixer.

Figure 27.2 shows two signals being added, but there is no real limit to the number of signals that could be added together. The mixer might have a dozen inputs, allowing you to simulate a violin chorus if you had enough function generators and filters.

At this point, however, your violin synthesis patch is not yet a satisfactory instrument. A real violin has a precise and flexible way of controlling the fundamental

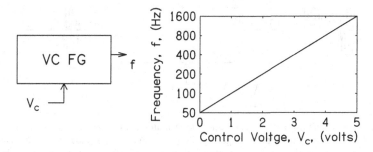

Fig. 27.3 The frequency of a voltage-controlled function generator (VCFG) depends on the control voltage inserted into the control port

Fig. 27.4 The frequency of the function generator (VCFG) is controlled both by the keyboard and by the low-frequency function generator (LFFG)

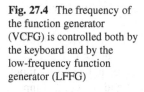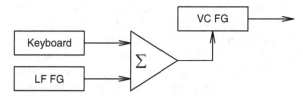

frequency of the tone. Precise control of the frequency is needed for melody, and the flexibility of the pitch is an essential part of artistic expression in violin playing. Your patch has none of that. It only has the knob on the front of the function generator to control the frequency. When you think about frequency control, you discover that you have real problems.

Because you are making a synthesizer, you would like to control the frequency with a keyboard. But you recall from Chap. 25 that the violin is played with vibrato, and therefore, the frequency control will need to incorporate the slow frequency modulation (FM). Evidently you need both kinds of control at once. To solve this problem you need to completely redesign the function generator. You need to make it voltage controlled. A voltage-controlled function generator (musicians often call it a voltage-controlled oscillator, or VCO) has a fundamental frequency that depends on a voltage that is put into a control port. If the voltage into the control port is low, the frequency is low. If the control voltage is high, the frequency is high (Fig. 27.3).

The voltage-controlled function generator allows a great deal of flexibility in pitch control because control voltages can be summed in a mixer, just as signal voltages were summed. Figure 27.4 shows how the frequency of the VCFG can be controlled in parallel by a keyboard (to play different notes) and by a low-frequency sine wave from another function generator (for vibrato). Voltages from the keyboard and from the low-frequency sine-wave function generator (LFFG) are summed to make a control voltage.

So far, so good. We can synthesize the tone of a violin (or two violins). We can control the frequency of the tone with a keyboard and give that tone realistic vibrato. There is a problem though—we cannot turn the tone on and off! As our

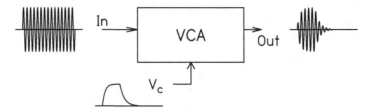

Fig. 27.5 The voltage-controlled amplifier (*VCA*) multiplies the input voltage by the control voltage (V_c) to produce an output voltage. Here, an input signal that is indefinitely long is turned on and off, and its amplitude is given a realistic shape by the envelope voltage (V_c) that is inserted into the control port for the VCA

patch of modules stands now, the tone is always on. We cannot just turn the tone on and off with a switch because such an abrupt onset and decay would not resemble the attack and decay of a violin tone. At this point we need a *voltage-controlled amplifier* (VCA) and a device called an envelope generator, as shown in Fig. 27.5.

The VCA has a signal input voltage, a signal output voltage, and a control input voltage. The signal output is the product of the signal input and the control input. Therefore, if the control input is 0 V, the output is zero. Now we can turn the tone off. If the control input rises in a gradual way and then dies away slowly, the output signal has a gradual attack and then fades out slowly. It is the role of the *envelope generator* to produce a slowly moving voltage that leads to a realistic attack and decay. The envelope begins its attack when the musician presses one of the keys of the keyboard.

Voltage control was a powerful idea in the early days of electronic music synthesizers because voltages can be easily and cheaply manipulated. They can be switched on and off, amplified, attenuated, inverted, summed, or distorted. A series of control voltages can be automated to produce a sequence of musical control gestures. Therefore, many different signal processing parameters were given voltage control. Most important were the voltage-controlled function generator (oscillator), the VCA, and voltage-controlled filter (where the cutoff frequency or passband frequency was determined by a control voltage). The voltage-controlled modules could be chained indefinitely. The vibrato example shown above was a short illustration of that.

The voltage-controlled analog synthesizer is no longer a mainstream instrument. It is awkward to use and expensive because it has a lot of front-panel hardware. Nevertheless, it is an important instrument because the functions of its voltage-controlled modules provided a common language and a useful framework for thinking about musical signal synthesis. These ideas are now implemented in digital synthesizers.

27.2 Digital Synthesizers

In a modern instrument, the modular functions are accomplished digitally. Signals, as such, do not exist in the digital synthesizer. Instead, signals are computed, and they remain in the form of digital numbers until finally converted to a voltage by a digital to analog converter. For example, while an analog synthesizer would add two voltages in a summer (perhaps two violin tones), the digital synthesizer adds two long sequences of numbers. Each sequence represents an instrument; each number represents an instantaneous value of the waveform produced by the instrument.

From the outset, digital methods are clearly promising. First, digital methods are fast. If there is a computational engine capable of millions of operations per second, then many operations can be done to generate each successive data point in the output. Second, digital methods can make use of computer memory. Memory can store both musical waveforms and control data, such as the times for turning notes on and off and the dynamics of a performance.

Speed and memory make a big difference. While an analog synthesizer is limited to one or two simultaneous voices, the speed of a digital synthesizer allows it to be truly polyphonic. The instrument can produce as many simultaneous notes as the musician can play on the keyboard. While an analog function generator is limited to a few simple waveforms, a digital memory can store waveforms digitally— perhaps recorded from real musical instruments. Pre-recorded real instrument sounds become useful when other synthesis techniques fail. For example, it is difficult to synthesize a convincing piano sound by any other means.

Because the digital synthesizer keeps all the music in memory before converting it to an analog voltage, it is automatically capable of performing signal processing techniques that add together multiple versions of the same signal with various delays. An application that immediately comes to mind is artificial reverberation. Without reverberation, synthesized music is weak. Artificial reverberation of some kind is always needed, and digital methods are ideal for this task. There are other applications of delay-and-add processing that came out of rock and roll studios. One is the technique called "chorusing" that thickens a musical voice by adding clones. Another is a technique called "flanging" that varies the amount of delay to produce a sonic effect described as "spacy."

The digital synthesizers that appear on stage at concerts are real-time performance instruments with a keyboard that can be played like a piano. The user can select from a few hundred different waveforms representing musical instruments or sound effects. Such sound "samples" are stored in waveform memory. Some synthesizers permit the user to record sounds into sample memory, to be played back later with different pitches. But the digital synthesizer is more than just a reproducer of sampled sounds. The recorded waveform samples can be processed as they are played back. Vibrato, filtering, spatial effects, and diverse forms of modulation can be added.

A digital synthesizer also includes a *sequencer*, which is a memory for musical gesture data. Such data can tell the signal generators to play the right notes at the right times with appropriate expression—level, level variation, vibrato, and pitch bend. The sequencer can store away enough gestural and performance data that it can play all the different parts of an entire piece of music by itself, once it is programmed to do so. Including the human musician in a live performance is still regarded as a nice touch, however. What is often done as a compromise is to preprogram rapid sequences of notes—arpeggios or riffs—so that pressing a particular key on the keyboard unleashes the entire sequence.

27.3   Musical Instrument Device Interface

The *musical instrument device interface*, or MIDI, is a data protocol that allows different electronic music instruments to communicate with one another. At a concert you will see a musician playing a keyboard instrument. It does not follow, however, that the instrument that he is playing is the instrument that is making the sounds! Quite possibly, the keyboard synthesizer is just being used as a control device to send gestural data to another synthesizer to make the sounds. The other synthesizer might be a box of electronics in a rack somewhere, or it might be a virtual synthesizer, which is a computer program running on a general purpose personal computer with a sound card.

The ability to send musical performance data around a stage or around a studio is made possible by the MIDI. Each musical device in the loop has a channel address, and data intended for that instrument are preceded with that channel number. Therefore, the MIDI connection requires only a single wire, running daisy-chain fashion among all the synthesizers or control devices. The intention of the MIDI protocol is to transfer all musically relevant information. There are set-up commands that tell a synthesizer to play a particular voice, such as a tom-tom or clarinet. There are performance data that turn notes on and off, and there are expressive data from keyboard pressure or pitch-bend wheels or foot pedals.

Because MIDI data are only control data, and not actual waveforms, MIDI data do not require a tremendous amount of memory and can be transmitted rapidly. MIDI data can be displayed on a computer monitor screen and can be edited to improve the performance. If you play a great riff except for one mistake, it is easy to correct the mistake with a computer editor and keep everything else.

The MIDI standard is an open standard, which means that any engineer can produce a musical instrument that reads and writes MIDI code. An important consequence is that different controllers can be created—replacing or augmenting the keyboard. Control technologies that resemble traditional musical instruments make it possible for many musicians to synthesize music digitally. It is even possible to synthesize sound based on a performer's brain waves, as measured by an electroencephalographic monitor, though the degree of control may not appeal to the performer or any other listener.

Science, Technology, Music, and Musicians

Electronic music is, in a sense, an ultimate in the application of technology to music. In one way or another it makes use of all the science to be found in this book. Making music electronically, either analog or digital, leads to an unprecedented opportunity for analysis and control of musical parameters.

In the early days of electronic music, people were impressed by the novelty of the sounds, and there was great interest in "what the electronic music synthesizer could do."

A perspective on this matter was given by Dr. Robert A. Moog (1934-2005), one of the pioneers of analog music synthesis. Bob Moog emphasized that at every period in history making music has involved the technology of the time. Ultimately, music was not made by the technology, it was made by musicians.

In October of 1974, Moog gave a talk at a meeting of the American Physical Society in Buffalo, New York. As he always did in his talks, he illustrated his points by playing compositions made on one of the synthesizers manufactured by his company. As I recall, he played a portion of *Snowflakes are Dancing* by Isao Tomita. At the conclusion, an enthusiastic member of the audience yelled out, "Oscillator, Oscillator!"

Without hesitation, Moog replied, "The next time I hear a violin concerto I am going to yell out, 'Cat, Cat!'" His point was that the electronic composition was no more played by the electronics than the violin concerto was played by catgut.

In the end, whatever the science or the technology, music is made by musicians. Musicians can take a couple of sticks and a rusty bucket, and use this "technology" to make music. The rational approach to musical sound adopted in *Principles of Musical Acoustics* is valuable intellectually. It may even help a musician better understand the principles of the craft. However, music surpasses the intellectual; it speaks to us on a deep and personal level. That is its real value.

Exercises

Exercise 1, Design this patch

The output of a voltage-controlled amplifier module (VCA) is given in terms of the input to the module by the following formula,

$$v_{out} = v_{in} \times (v_c/5) \qquad (v_c \geq 0) \qquad (1)$$

$$v_{out} = 0 \qquad\qquad (v_c < 0) \qquad\qquad\qquad (2)$$

where v_c is the control voltage in volts. Thus when the control voltage is zero, there is no output. When the control voltage is 5 V, the output is equal to the input.

Draw a patch including two function generators and a VCA to generate an amplitude-modulated signal, as described in Chap. 21.

Exercise 2, Digital speed

Imagine a digital synthesizer that performs 100,000,000 operations per second. It generates a signal at a sample rate of 44,100 samples per second, the sample rate of the compact disc. Show that this synthesizer can perform more than 2,200 operations for every output sample.

Exercise 3, Emulation

As a relatively new art form, electronic music is still trying to decide what it wants to be. Recently designers have worked hard to synthesize the sounds of traditional acoustical instruments. Others say that this is a mistake. They think that the role of electronic music should be to create new sounds. What do you think?

Exercise 4, Wheels

Many synthesizers have two controllers shaped like wheels to be controlled with the musician's left hand. (a) One of them controls the depth of the vibrato. (b) The other is able to bend the pitch of a note. Why would one go to the trouble of providing such elaborate frequency control?

Exercise 5, Computer application

Is a general purpose computer, like a PC, more naturally a sequencer or a synthesizer?

♠

Appendix A: Composers

Renaissance

Giovanni da Palestrina 1525(?)–1594
Josquin des Pres c 1450–1521
Thomas Tallis 1505–1585
Michael Praetorius 1571–1621
Claudio Monteverdi 1567–1643

Baroque

Dietrich Buxtehude 1637–1707
Johann Pachelbel 1653–1706
Arcangelo Corelli 1653–1713
Henry Purcell 1659–1695
Antonio Vivaldi 1678–1741
Georg Philipp Telemann 1681–1767
Jean-Philipe Rameau 1683–1764
Johann Sebastian Bach 1685–1750
Georg Frederick Handel 1685–1759
Giovani Pergolesi 1710–1736
Christoph W. Gluck 1714–1787

Rococo

Francois Couperin 1668–1733
Domenico Scarlatti 1685–1757

Wilhelm Friedman Bach 1710–1784
Carl Phillip Emanuel Bach 1714–1788
PDQ Bach 1807–1742 (?)

Classic

Josef Haydn 1732–1809
Johann Christian Bach 1735–1782
Luigim Boccherini 1743–1805
Muzio Clementi 1752–1832
Wolfgang Amadeus Mozart 1756–1791
Antonio Salieri 1750–1825

Romantic

Ludwig van Beethoven 1770–1827
Gioacchino Rossini 1792–1868
Franz Schubert 1797–1828
Hector Berlioz 1803–1869
Felix Mendelssohn 1809–1847
Robert Schumann 1810–1856
Richard Wagner 1813–1883
Johannes Brahms 1833–1897
Cesar Franck 1822–1890
Anton Bruckner 1824–1896
Camille Saint-Saens 1835–1921
Pablo de Sarasate 1884–1908

W.M. Hartmann, *Principles of Musical Acoustics*, Undergraduate Lecture Notes
in Physics, DOI 10.1007/978-1-4614-6786-1,
© Springer Science+Business Media New York 2013

Virtuoso

Niccolo Paganini 1782–1840
Frederic Chopin 1810–1849
Franz Liszt 1811–1886

Russian Movement

Alexander Borodin 1834–1887
Modest Musorgsky 1839–1881
Petr I. Tchaikovsky 1840–1893
Nicholai Rimsky-Korsakov 1844–1908
Alexander Scriabin 1872–1915
Serge Prokofiev 1891–1952

Nationalist

Bedrich Smetana 1824–1884
Antonin Dvorak 1841–1904
Edvard Grieg 1843–1907
Carl Nielsen 1865–1931
Jean Sibelius 1865–1957

Impressionists

Claude Debussy 1862–1918
Fredrick Delius 1862–1934
Eric Satie 1866–1925
Maurice Ravel 1875–1937
Ottorino Respighi 1879–1936
Charles T. Griffes 1845–1920

Post Romantics

Edward Elgar 1857–1934
Gustav Mahler 1860–1911
Richard Strauss 1864–1949

Ralph Vaughn Williams 1872–1958
Sergei V. Rachmaninoff 1873–1943
Gustav Holst 1874–1934

Expressionists–Serialists

Arnold Schoenberg 1874–1951
Anton von Webern 1883–1945
Alban Berg 1885–1935

Neo Classic

Charles Ives 1874–1954
Manuel de Falla 1876–1946
Darius Milhaud 1892–1974
Walter Piston 1894–1976
Carl Orff 1895–1982
Howard Hansen 1896
Roger Sessions 1896–1985
Francis Poulenc 1899–1963
Joaquin Rodrigo 1902–1999
Aaron Copland 1906–1990
Dmitri Shostakovich 1906–1975
Samuel Barber 1910–1981
William Schuman 1910–1992
Benjamin Britten 1913–1976

Modern

Bela Bartok 1881–1945
Igor Stravinsky 1882–1971
Edgar Varese 1885–1965
Paul Hindemith 1895–1963
Harry Partch 1901–1974
Olivier Messiaen 1908–1992
John Cage 1912–1992
Pierre Boulez 1925–
Luciano Berio 1925–2003
Karlheinz Stockhausen 1928–2007

Appendix B: Trigonometric Functions

Definitions

The trigonometric functions are defined in terms of a right triangle, shown in Fig. B1.

Figure B1 shows angle θ in the first quadrant of the x–y plane. The adjacent and opposite sides are both positive numbers in this quadrant. Further, the hypotenuse is always a positive number. Therefore, both sine and cosine functions are positive in this quadrant. Figure B2 shows angle θ in other quadrants and gives the signs for sine and cosine functions.

By definition, the hypotenuse (with length hyp) is the side opposite to the right angle (90°) and it is the longest side. The angle of interest (θ) determines the definition of the "opposite side" (op) and the "adjacent side" (adj). The basic trigonometric functions are the ratios of the lengths of these sides. Because they are ratios, the trigonometric functions depend only upon angle θ; they do not depend upon the size of the triangle:

$$\sin \theta = (\text{op})/(\text{hyp}) \qquad \text{The sine function.} \qquad \text{(B1)}$$

$$\cos \theta = (\text{adj})/(\text{hyp}) \qquad \text{The cosine function.} \qquad \text{(B2)}$$

$$\tan \theta = (\text{op})/(\text{adj}) \qquad \text{The tangent function.} \qquad \text{(B3)}$$

Using Fig. B1 and the definitions of the sine, cosine, and tangent, we see that the sine and cosine functions can only have values from +1 to −1. The tangent function, however, can have any value between plus and minus infinity.

The definitions immediately relate the tangent function to the sine and cosine functions:

$$\tan \theta = \frac{\sin \theta}{\cos \theta} \qquad \text{for all } \theta. \qquad \text{(B4)}$$

The Pythagorean relation for a right triangle relates the lengths of the sides:

W.M. Hartmann, *Principles of Musical Acoustics*, Undergraduate Lecture Notes in Physics, DOI 10.1007/978-1-4614-6786-1,
© Springer Science+Business Media New York 2013

Fig. B1 An angle in the first quadrant. The opposite side and the adjacent side are both positive

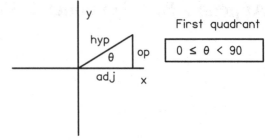

Fig. B2 Angles in the other three quadrants

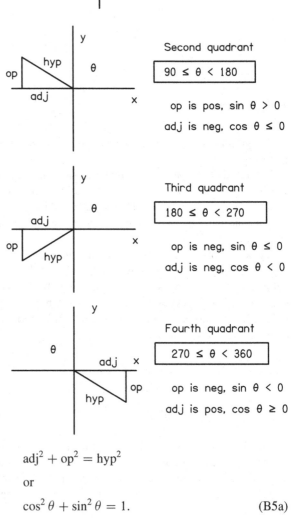

$$adj^2 + op^2 = hyp^2$$

or

$$\cos^2 \theta + \sin^2 \theta = 1. \tag{B5a}$$

♠

Appendix C: Traditional Musical Instruments

This appendix indicates the fundamental frequencies in units of Hertz (Hz) of significant notes on some important instruments of the orchestra.

For most instruments the bottom note (lowest frequency) is rather firmly fixed by the construction of the instrument, though non-traditional playing techniques may produce lower frequencies. The top notes are much less well defined. For instance, the top note on the trumpet is given as 1,244 Hz, but some jazz players can hit the "double high C" at $B\flat_6 = 1,863$ Hz. Normally the top notes given below are those that a musician may expect to see written in musical scores.

It is important to remember that the entries below only correspond to the *fundamental* frequencies of instruments. Each of the tones has harmonics that lead to energy at much higher frequencies. For instance, the bottom note of the trumpet is listed as 165 Hz, but there is lots of energy in the 10th harmonic of this note, and that frequency is 1,650 Hz (higher than the fundamental frequency of the top note).

All frequencies have been rounded off to the nearest Hz. Careful tuning tries to be more precise than this.

Strings

There are four or six strings. Their note names are given first in order of ascending frequency. Then the corresponding fundamental frequencies are given.

Violin: G_3, D_4, A_4, E_5, 196, 294, 440, 659, top at $E_7 = 2637$
Viola: C_3, G_3, D_4, A_4, 131, 196, 294, 440, top at $A_6 = 1760$
Cello: C_2, G_2, D_3, A_3, 65, 98, 147, 220, top at $E_5 = 659$
Bass: E_1, A_1, D_2, G_2, 41, 55, 73, 98, top at $D_4 = 587$
Guitar: E_2, A_2, D_3, G_3, B_3, E_4, 82, 110, 147, 196, 247, 330, top at $B_5 = 938$

W.M. Hartmann, *Principles of Musical Acoustics*, Undergraduate Lecture Notes in Physics, DOI 10.1007/978-1-4614-6786-1,
© Springer Science+Business Media New York 2013

Woodwinds

Flute: bottom at $C_4 = 262$, top at $D_7 = 2349$
Piccolo: bottom at $C_5 = 523$, top at $C_8 = 4186$
Clarinet*: bottom at $D_3 = 147$, top at $B\flat_6 = 1865$
Alto Saxophone*: bottom at $D\flat_3 = 139$, top at $A\flat_5 = 831$
Tenor Saxophone*: bottom at $A\flat_2 = 104$, top at $E\flat_5 = 622$
Oboe: bottom at $B\flat_3 = 233$, top at $G_6 = 1568$
Bassoon: bottom at $B\flat_1 = 58$, top at $E\flat_5 = 622$

Brass

Trumpet*: bottom at $E_3 = 165$, top at $E\flat_6 = 1244$
Trombone*: bottom at $E_2 = 82$, top at $D_5 = 587$
Tuba*: bottom at $E\flat_1 = 39$, top at $B\flat_3 = 233$
French horn*: bottom at $B_1 = 62$, top at $F_5 = 698$

Other

Tenor voice: top (not falsetto) at approximately $C_5 = 523$
Harp: bottom at $B_0 = 31$, top at $A_7 = 3520$
Piano: bottom at $A_0 = 28$, top at $C_8 = 4186$
Organ: bottom at $C_0 = 16$, top at $C_9 = 8372$

*This table gives frequencies and note names that are, so-called, "concert" notes, established by the notes on the piano or the notes recognized by a violin player. An instrument marked with an asterisk is a "transposing" instrument, meaning that the note name recognized by the player is different from the "concert" note name. For instance, the trumpet is a $B\flat$ instrument. Therefore, when the trumpet player plays what he reads on the score as "middle C," and what he calls "middle C," the fundamental frequency of the note that sounds is not C-262 Hz. Instead it is $B\flat$-233 Hz. The note $B\flat$ is a whole step lower than C. The transposition is consistent in that the notes written for a trumpet always sound a whole step lower than concert pitch.

♠

Appendix D: Keyboard Tuning

Keyboard instruments like pianos, organs, and synthesizers give a musician the opportunity to play many notes at once. In exchange for this benefit, the keyboard player sacrifices control of intonation—fine control of the frequency during playing. The piano player cannot bend notes up or down like the player of wind instruments or string instruments. The individual notes of a keyboard instrument must be tuned before the instrument is played.

Figure D1 shows a standard keyboard with notes named C, $C\sharp$, D, etc. It also shows musical intervals, unison, minor second, Major second, etc., using C as a reference. The figure shows a little more than an octave. You will note that the pattern of the keys is the same for each octave. An octave corresponds to a factor of two—*always*, whatever the tuning. For instance, if the note called C on the far left has a frequency of 261.6 Hz then the note called C to the right (the "Octave") has a frequency of $261.2 \times 2 = 522.4$ Hz.

Table D1 shows frequency ratios corresponding to those intervals for three different tunings. The table gives the frequency ratios for musical intervals within the octave in three tunings, Just, Pythagorean, and Equal Tempered, per ANSI standard S1.1-1994. There are 12 intervals besides the unison.

Just Tuning The tuning called "Just" chooses the frequency ratios to be the ratios of small integers like 5/4 for the interval called the "Major third." For instance, if we tune a keyboard in the key of C and make the frequency of middle C equal to 261.6 Hz, then the frequency of the note called E (Major third) will be $261.6 \times 5/4 = 327.0$ Hz. The Just tuning leads to very smooth sounding chords.

Pythagorean Tuning The tuning called "Pythagorean" is obtained by using the ratios 3/2 and 4/3 systematically. For example the weird ratio 1024/729 is equal to 4/3 raised to the sixth power $(4/3)^6 = 4096/729$, which is then brought into the first octave by twice dividing by 2 (i.e., by 4). For instance, if we tune a keyboard in the key of C and make the frequency of middle C equal to 261.6 Hz, then the frequency of the note called E (Major third) will be $261.6 \times 81/64 = 331.1$ Hz.

W.M. Hartmann, *Principles of Musical Acoustics*, Undergraduate Lecture Notes
in Physics, DOI 10.1007/978-1-4614-6786-1,
© Springer Science+Business Media New York 2013

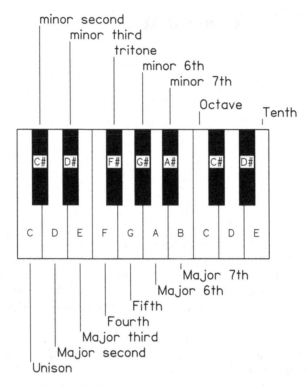

Fig. D1 A section of the piano keyboard with note names and interval names. The assignment of interval names assumes that the low C is the reference

The Pythagorean tuning was the starting point for many different tunings of the baroque era known as meantone tunings.

Equal Temperament The tuning called "Equal temperament" chooses the frequency ratios to be powers of $2^{1/12}$, the ratio of the semitone. Therefore, any two adjacent notes on the keyboard have a frequency ratio of $2^{1/12}$, regardless of the key. For instance, whatever key the music may be, if we make the frequency of middle C equal to 261.6 Hz, then the frequency of the note called E (Major third) will be $261.6 \times 2^{4/12} = 329.6$ Hz. Equal temperament is used almost universally in our own day. It has all but replaced the other tunings.

Table D1 Notes of the scale in three temperaments

Interval	Just tuning	Pythagorean tuning	Equal temperament
Unison	1.	1.	$2^0 = 1.$
minor Second	$16/15 = 1.067$	$256/243 = 1.053$	$2^{1/12} = 1.059$
Major Second	$10/9 = 1.111$ or $9/8 = 1.125$	$9/8 = 1.125$	$2^{2/12} = 1.122$
minor Third	$6/5 = 1.200$	$32/27 = 1.185$	$2^{3/12} = 1.189$
Major Third	$5/4 = 1.250$	$81/64 = 1.266$	$2^{4/12} = 1.260$
Fourth	$4/3 = 1.333$	$4/3 = 1.333$	$2^{5/12} = 1.335$
Tritone	$45/32 = 1.406$ or $64/45 = 1.422$	$1024/729 = 1.405$ or $729/512 = 1.424$	$2^{6/12} = 1.414$
Fifth	$3/2 = 1.500$	$3/2 = 1.500$	$2^{7/12} = 1.498$
minor Sixth	$8/5 = 1.600$	$128/81 = 1.580$	$2^{8/12} = 1.587$
Major Sixth	$5/3 = 1.667$	$27/16 = 1.688$	$2^{9/12} = 1.682$
minor Seventh	$7/4 = 1.750$ or $16/9 = 1.778$ or $9/5 = 1.800$	$16/9 = 1.778$	$2^{10/12} = 1.782$
Major Seventh	$15/8 = 1.875$	$243/128 = 1.898$	$2^{11/12} = 1.888$
Octave	$2/1 = 2.000$	$2/1 = 2.000$	$2^{12/12} = 2.000$

♠

Appendix E: Standard Musical Frequencies

This appendix gives the frequencies (in Hertz) of the notes of the equal-tempered musical scale assuming that concert A is tuned to 440 Hz, per the international standard.

Note	Freq	Note	Freq	Note	Freq
C_0	16.35	C_2	65.41	C_4	261.6
$C\sharp_0$ or Db_0	17.32	$C\sharp_2$ or Db_2	69.30	$C\sharp_4$ or Db_4	277.2
D_0	18.35	D_2	73.42	D_4	293.7
$D\sharp_0$ or Eb_0	19.45	$D\sharp_2$ or Eb_2	77.78	$D\sharp_4$ or Eb_4	311.1
E_0	20.60	E_2	82.41	E_4	329.6
F_0	21.83	F_2	87.31	F_4	349.2
$F\sharp_0$ or Gb_0	23.13	$F\sharp_2$ or Gb_2	92.50	$F\sharp_4$ or Gb_4	370.0
G_0	24.50	G_2	98.00	G_4	392.0
$G\sharp_0$ or Ab_0	25.96	$G\sharp_2$ or Ab_2	103.83	$G\sharp_4$ or Ab_4	415.3
A_0	27.50	A_2	110.00	A_4	440.0
$A\sharp_0$ or Bb_0	29.14	$A\sharp_2$ or Bb_2	116.54	$A\sharp_4$ or Bb_4	466.2
B_0	30.87	B_2	123.47	B_4	493.9
C_1	32.70	C_3	130.8	C_5	523.3
$C\sharp_1$ or Db_1	34.65	$C\sharp_3$ or Db_3	138.6	$C\sharp_5$ or Db_5	554.4
D_1	36.71	D_3	146.8	D_5	587.3
$D\sharp_1$ or Eb_1	38.89	$D\sharp_3$ or Eb_3	155.6	$D\sharp_5$ or Eb_5	622.3
E_1	41.20	E_3	164.8	E_5	659.3
F_1	43.65	F_3	174.6	F_5	698.5
$F\sharp_1$ or Gb_1	46.25	$F\sharp_3$ or Gb_3	185.0	$F\sharp_5$ or Gb_5	740.0
G_1	49.00	G_3	196.0	G_5	784.0
$G\sharp_1$ or Ab_1	51.91	$G\sharp_3$ or Ab_3	207.7	$G\sharp_5$ or Ab_5	830.6
A_1	55.00	A_3	220.0	A_5	880.0
$A\sharp_1$ or Bb_1	58.27	$A\sharp_3$ or Bb_3	233.1	$A\sharp_5$ or Bb_5	932.3
B_1	61.74	B_3	246.9	B_5	937.8

W.M. Hartmann, *Principles of Musical Acoustics*, Undergraduate Lecture Notes in Physics, DOI 10.1007/978-1-4614-6786-1,
© Springer Science+Business Media New York 2013

C_6	1046.5	C_7	2093.	C_8	4186.
$C\sharp_6$ or Db_6	1108.7	$C\sharp_7$ or Db_7	2218.	$C\sharp_8$ or Db_8	4435.
D_6	1174.7	D_7	2349.	D_8	4699.
$D\sharp_6$ or Eb_6	1244.5	$D\sharp_7$ or Eb_7	2489.	$D\sharp_8$ or Eb_8	4978.
E_6	1318.5	E_7	2637.	E_8	5274.
F_6	1396.9	F_7	2794.	F_8	5588.
$F\sharp_6$ or Gb_6	1480.0	$F\sharp_7$ or Gb_7	2960.	$F\sharp_8$ or Gb_8	5920.
G_6	1568.0	G_7	3136.	G_8	6272.
$G\sharp_6$ or Ab_6	1661.2	$G\sharp_7$ or Ab_7	3322.	$G\sharp_8$ or Ab_8	6645.
A_6	1760.0	A_7	3520.	A_8	7040.
$A\sharp_6$ or Bb_6	1864.6	$A\sharp_7$ or Bb_7	3729.	$A\sharp_8$ or Bb_8	7459.
B_6	1975.5	B_7	3951.	B_8	7902.

Appendix F: Power Law Dependences

Linear (First Power) Dependence If one pencil costs 10 cents, then two pencils costs 20 cents, and three pencils costs 30 cents, etc. The total cost of pencils depends linearly on the number of pencils purchased. We say that the cost varies as the *first* power of the number of pencils. Double the number of pencils and you double the cost.

Square Law (Second Power) Dependence If the side of a square has a length of 5 in., then the square has an area of 25 in.^{2}. If the side of a square has a length of 10 in., then the square has an area of 100 in.^{2}. The area depends on the *square* of the length of a side. We say that the area varies as the *second* power of the length. Double the length of a side and you multiply the area by four.

The linear law and the square law are examples of power laws. When there is a power law, it is possible to make statements about how many times larger or smaller a dependent quantity becomes when another quantity is varied.

Here is another power law, based on Eq. (3.1). That equation says that the frequency of a simple harmonic vibration varies as the square root of the stiffness, quantity s. To gain insight into how the equation works, we ask what happens if the stiffness is doubled. We will answer this question in two ways, one quick, the other slow and systematic. We reproduce Eq. (3.1) for convenient reference

$$ f = \frac{1}{2\pi} \sqrt{\frac{s}{m}} \tag{3.1} $$

Quick Thinking: Square-Root Law (One-Half Power) Dependence Because the frequency varies as the square root of the stiffness, if we double the stiffness we multiply the frequency by the square root of 2. For instance, if the frequency is initially 100 Hz then after the stiffness is doubled, the frequency becomes $100 \times \sqrt{2}$, or 141 Hz. That is the quick way.

W.M. Hartmann, *Principles of Musical Acoustics*, Undergraduate Lecture Notes in Physics, DOI 10.1007/978-1-4614-6786-1, 299
© Springer Science+Business Media New York 2013

Slow and Systematic We identify two conditions:

OLD CONDITION: Stiffness s gives frequency of 100 Hz.

NEW CONDITION: Stiffness $2s$ gives a new, unknown frequency.

Our task is to find the new frequency.

We write Eq. (3.1) for both conditions

$$f_{OLD} = \frac{1}{2\pi}\sqrt{s_{OLD}/m_{OLD}}$$

$$f_{NEW} = \frac{1}{2\pi}\sqrt{s_{NEW}/m_{NEW}}$$

We can divide the new equation by the old equation:

$$\frac{f_{NEW}}{f_{OLD}} = \frac{\frac{1}{2\pi}\sqrt{s_{NEW}/m_{NEW}}}{\frac{1}{2\pi}\sqrt{s_{OLD}/m_{OLD}}}.$$

The factors of 2π appear in both the numerator and the denominator, and they cancel. Likewise, the old mass and the new mass are the same and they too cancel. Therefore,

$$\frac{f_{NEW}}{f_{OLD}} = \frac{\sqrt{s_{NEW}}}{\sqrt{s_{OLD}}}.$$

Because the ratio of square roots is equivalent to the square root of the ratios,

$$\frac{f_{NEW}}{f_{OLD}} = \sqrt{\frac{s_{NEW}}{s_{OLD}}}.$$

We have succeeded in reducing this equation down to something we know, because we know that the new stiffness is twice the old stiffness. Therefore,

$$\frac{f_{NEW}}{f_{OLD}} = \sqrt{2}$$

or

$$f_{NEW} = f_{OLD}\sqrt{2}.$$

Now if f_{OLD} is 100 Hz then f_{NEW} is $100\sqrt{2} = 141$ Hz.

That is the end of the slow and systematic way, and it gives the same answer as the quick way. In the quick way we just look at the dependence and apply it directly. For instance, if the stiffness is increased by a factor of three (becomes three times larger), then the frequency is increased by the square root of 3. If it was originally 100 Hz, then after increase the frequency f_{NEW} is $100\sqrt{3} = 173\,$Hz.

Inverse Square-Root Law (Minus One-Half Power) Dependence Equation (3.1) shows that the frequency depends inversely on the square root of the mass. "Inverse" means reciprocal. For instance, if the mass is doubled then the frequency becomes smaller. It becomes smaller by a factor of the reciprocal of the square root of two. That means that $f_{NEW} = f_{OLD}/\sqrt{2}$. If the frequency was originally 100 Hz, then after the increase in mass the frequency f_{NEW} is $100/\sqrt{2} = 71\,$Hz.

♠

Appendix G: Telephone Tones

The telephone uses pairs of sine tones to encode the numbers in touch-tone dialing. The sine tones are added together with approximately equal amplitudes. The sine tone frequencies (in Hz) are given in the table below

Frequencies (Hz)	1209	1336	1477	1633
697	1	2	3	(not
770	4	5	6	yet
852	7	8	9	used)
941	*	0	#	

For example pressing the button called "4" sends a signal with components at 770 and 1,209 Hz.

The dial tone of a telephone is also a sum of sine tones, with frequencies 350 and 440 Hz.

The dial tone is different from the tones from the numeric keypad because it is consonant. The ratio 440/350 equals 1.257, which is an interval of a Major third. For comparison note that Appendix D gives the Major third as 1.25 in Just tuning, 1.266 in Pythagorean, and 1.260 in Equal temperament. Historically, there have been so many different meantone tunings with different ratios for the Major third, that the ratio 1.257 must have appeared in somebody's favorite tuning.

The telephone bandwidth for speech is 300 Hz to 3.3 kHz. The dynamic range is about 40 dB.

♠

W.M. Hartmann, *Principles of Musical Acoustics*, Undergraduate Lecture Notes in Physics, DOI 10.1007/978-1-4614-6786-1,
© Springer Science+Business Media New York 2013

Appendix H: Greek Alphabet

Some of the equations in this book use letters of the Greek alphabet. Here are all 24 of them, upper case, lower case, and their names.

A	α	Alpha
B	β	Beta
Γ	γ	Gamma
Δ	δ	Delta
E	ϵ	Epsilon
Z	ζ	Zeta
H	η	Eta
Θ	θ	Theta
I	ι	Iota
K	κ	Kappa
Λ	λ	Lambda
M	μ	Mu
N	ν	Nu
Ξ	ξ	Xi
O	o	Omicron
Π	π	Pi
P	ρ	Rho
Σ	σ	Sigma
T	τ	Tau
Υ	υ	Upsilon
Φ	ϕ	Phi
X	χ	Chi
Ψ	ψ	Psi
Ω	ω	Omega

♠

W.M. Hartmann, *Principles of Musical Acoustics*, Undergraduate Lecture Notes in Physics, DOI 10.1007/978-1-4614-6786-1,
© Springer Science+Business Media New York 2013

Answers to Exercises

Chapter 1

1.1: A reflecting wall is part of the transmission path. A musical instrument is a source. A microphone is a receiver of acoustical waves, but in turn, it transmits electrical waves. So it could be considered either a receiver or part of the transmission path.

1.2: Fish perceive sound vibrations in the water, whales communicate, and dolphins and submarines echolocate, all with sound waves. Sounds travel very well in solids. It is believed that elephants hear waves that are transmitted through the ground. Putting your ear on a railroad track can give the first evidence that a train is coming. Mostly, however, the sounds that are transmitted through solids, such as a solid wall, are an annoyance.

1.3: If a sound source is far away it may be too weak to hear. Intense sounds can *mask* quieter sounds, such as the sounds of a conversation that you want to hear. Excess reverberation in a room can cause speech to be garbled. Listening to a single talker in a context with many other talkers is often possible, but stressful.

1.5: Ocean waves and X-rays would be examples.

1.6: The difference is $137 - 120 = 17\,\text{kg}$. The ratio is $137/120 = 1.14$. The ratio has no units. It would be wrong to say that the ratio is $1.14\,\text{kg}$. The percentage change is $100 \times (137/120 - 1) = 14\,\%$.

1.7: The simple way to do this problem is to use the idea of Eq. (1.2) to realize that decreasing a number by $10\,\%$ is equivalent multiplying it by 0.9, and that increasing a number by $10\,\%$ is equivalent to multiplying it by 1.1. Thus decreasing and then increasing by $10\,\%$ multiplies the original number by $0.9 \times 1.1 = 0.99$ or $99\,\%$. Hence the change is a $1\,\%$ loss. The order of decreasing and increasing events does not matter because the order of multiplying two numbers together does not matter.

W.M. Hartmann, *Principles of Musical Acoustics*, Undergraduate Lecture Notes in Physics, DOI 10.1007/978-1-4614-6786-1,
© Springer Science+Business Media New York 2013

The long way to do the problem is to take a particular case. Start with $100. Watch it decrease by 10 % to $90 and then watch it increase by 10 % to $99. Then do the reverse. Start with $100. Watch it increase by 10 % to $110 and then watch it decrease by 10 % to $99.

1.8: (a) 10^5, (b) 10, (c) 10, (d) 10^5, (e) 1100.

1.9: (a) $ 2.0 × 10^{10}$, (b) $ 1.0 × 10^{-1}$, (c) 2.31 × 10^5, (d) 3.4 × 10^{-4}.

1.10: (a), (b), and (c) One million for all three.

Chapter 2

2.1: 1/20 s = 0.05 s = 50 ms. 1/20,000 s = 0.05 ms or 50 μs.

2.2: 1/60 = 0.01667 Hz.

2.3: (a) 2 s = 2,000 ms; (b) 30 μs = 0.03 ms.

2.4: (a) 16,384 Hz = 16.384 kHz. (b) 10 kHz = 10,000 Hz.

2.5: (a) $f = 1/T = 1/0.001 = 1,000$ Hz = 1 kHz
   (b) and (c) $T = 1/f = (1/10,000)$ s = 0.1 ms = 100 μs.

2.6: In Fig. 2.4 the dashed line is a reference, and its phase ϕ is zero. Compared to this reference another wave is lagging if its phase angle is between 0 and $-180°$.

2.7: "Infrared" light has a frequency that is too low to see. The term "infrared" means light with a frequency lower than red light—the lowest frequency of light we can see.

2.8: The band from 300 to 3,300 Hz comprises three octaves. You can prove that by starting at 300 and progressing by octaves: up one octave is 600, up two octaves is 1,200, up three octaves is 2,400. To go up a fourth octave leads to 4,800 Hz, but that is outside the band. Therefore, the telephone bandwidth is between three and four octaves.

2.9:

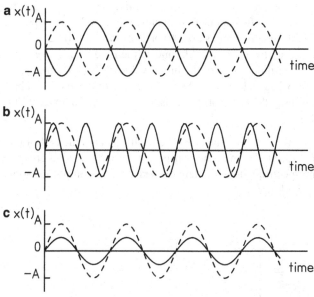

For Exercise 2.9 (a–c). The *dashed line* is the reference sine

2.9 (d): There are many possible answers to this question. Almost any scribble that a child would do is likely to be a function that is not single-valued. Figure below shows two simple functions. You can see that in each case knowing the value of x could leave you uncertain about the value of y.

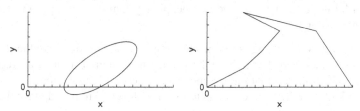

For Exercise 2.9 (d). Two functions that are not single-valued

Chapter 3

3.1: The square root of (100/0.25) is 20. Divide by 2π to get 3.18 Hz.

3.2: The mass is near the ends of the tines, and the stiffness comes from the shoulders. File the tines and the frequency increases because the mass gets smaller. File the shoulders and the frequency decreases because with less material there is less stiffness.

3.3: In engineering, one measures the lifetime as the time it takes for the amplitude to fall below its initial value by a factor of $1/e$, where e is the natural logarithm base, $e \approx 2.72$. That corresponds to 37 % of the initial value. In Fig. 3.1, the amplitude has dropped to 37 % after about four cycles. Each cycle has a period of 1/200 s. Therefore, the lifetime is 4 times (1/200) or 1/50, which is 0.02 s.

In acoustics, one measures the lifetime as the time it takes for the amplitude to fall below one one-thousandth of the initial value (0.001). By the that standard, the lifetime in Fig. 3.1 is at least as long as the entire time axis with about 18 cycles. Therefore, we'll say the waveform has disappeared after 18 times (1/200) or 18/200, which is 0.09 s.

3.4: The recipe for salt is as follows: "Add salt." That is the end of the recipe. The spectrum of a 1,000-Hz sine tone consists of a 1,000-Hz sine component. That is the end of the spectrum. The problem does not state an amplitude so we are free to pick any amplitude we like.

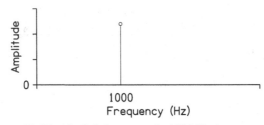

For Exercise 3.4. Spectrum of a 1,000-Hz sine tone

3.6: There are two people (systems) and a suggestion made by one of them is gladly followed by the other because the two people (systems) are in agreement on the matter in question.

3.7: (a) A child on a swing is like a swinging pendulum. The pendulum has one mode of vibration and if you drive the pendulum system at its natural frequency the amplitude of the oscillation becomes huge. Thus pushing a child in a swing is just like adding energy to a pendulum to increase the amplitude and make the child swing higher.
(b) The playground swing is not an exact example of resonance. The driving force is not a sine wave force. Instead, it is impulsive. If you push the child on every other cycle of motion, or every third, the amplitude increases, even though the driving frequency is one-half or one-third of the natural frequency of vibration. Strictly speaking, resonance is not like that. The driving frequency and the mode frequency have to be exactly the same for resonance.

3.8: (a) The frequency stays the same. (b) The frequency gets smaller by a factor of the square root of two. That means that if the frequency is originally f_1 then after doubling the mass the frequency becomes $f_1/\sqrt{2}$ or $f_1/1.414$.

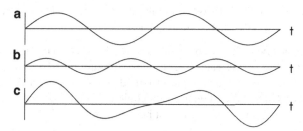

For Exercise 3.9. Adding (a) and (b) point-by-point gives you (c)

3.9: (a)

(b) For every cycle of *a* there are 1.5 cycles of *b*. Therefore, the frequency of *b* is 1.5 times the frequency of *a*. If the frequency of *a* is 100 Hz the frequency of *b* is 150 Hz.

3.10: None of these is an example of resonance because the driving force is not a sine function and so the driving system does not have any particular frequency. The essential condition for resonance is that the frequency of the driving must equal a natural frequency of the driven system.

3.11: The top pattern shows the mode with the highest frequency and the bottom pattern shows the lowest. Complicated vibration patterns generally have high frequencies.

3.12: A larger volume leads to lower frequencies for the modes of vibration of air in the box. Lower frequency modes will resonate with tines of lower pitch.

Chapter 4

4.1: There are two transducers, a microphone and a miniature loudspeaker.

4.2: Many of different periodic waveforms might have two positive-going zero crossings per cycle. Here is one of them

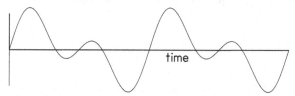

For Exercise 4.2. A waveform to fool the frequency counter

The frequency counter adds a count to the total whenever the signal input circuit detects a positive-going zero crossing. With two of these per cycle, the total count, after any span of time, will be twice the number of cycles. Therefore, the counter will read twice the true fundamental frequency.

4.3: If the frequency really is 262.3 Hz then in 10 s there will be 2,623 positive-going zero crossings. If the gate is set to have a duration of 10 s, then the counter will register 2,623 counts. One only has to divide by ten to get an average of 262.3 counts per second. The down side to this technique is that the measurement takes a full 10 s. During that time the signal must be completely stable.

4.4: Guitar pickups, headphones, tape heads, compact disc read and write heads are optical transducers. A light bulb is also a transducer—it converts electrical energy into heat and light energy. The list goes on and on.

4.5: For a 500-Hz wave the period is 2 ms. To display two periods in a single sweep requires that the sweep be 4 ms long. The sweep period should be 4 ms. Therefore, the sawtooth wave that drives the horizontal deflection mechanism should have a rate of 1/(4 ms) or 250 Hz.

4.6: When it comes to comparisons, two is a special number. If you can compare two things then you can compare an indefinitely large number of things by working serially, two at a time. The dual-trace oscilloscope allows you to compare waveforms A and B. If you then compare waveform A with waveform C, you can infer the relationship between C and B. You can continue by comparing A with D and so on, without end. By contrast, a single-trace scope does not allow simultaneous comparisons.

4.7: You want the electron attracted to the left plate and be repelled from the right plate, so that it hits the left side of the oscilloscope screen. Thus since the electron has a negative charge the left plate should be positively charged, and the right plate should be negatively charged.

4.8: The answer is 800 Hz. Here is how you get it. 10 divisions × 0.5 ms/division = 5 ms for the entire screen width. 4 (cycles/screen)/5 (ms/screen) = 4 cycles/(5 ms) or 0.8 cycles/ms or 800 cycles/s, and so 800 Hz.

4.9: Two divisions would correspond to 2 × 0.5 ms or 1 ms. The period is slightly longer than 1 ms. Therefore, the frequency is slightly less than 1,000 Hz.

4.11: 0.5 v/cm vertically and 1 ms/cm (or 0.001 s/cm) horizontally.

Chapter 5

5.1: It is not the constant part of the pressure that makes a sound painfully loud. In fact, the first 0.2 ms of the sound in the figure have a pressure of 101,325 Pa, and there is no sound at all. It is the variation in pressure that causes a sound, and a large variation can create a sound that is so loud it is painful.

5.2: Remember that $1\,m^2$ is $10,000\,cm^2$. Therefore, to find the pressure in Newtons per square centimeter you divide by 10,000.

It is useful to apply the concepts of unit analysis to this exercise. The exercise requires that we convert from units of Newtons per square meter to units of Newtons per square centimeter. To do this operation we need to multiply by something that is equivalent to 1.0 that gets rid of the square meters and replaces it with square centimeters. The kind of thing that is likely to work is a fraction with square meters in the numerator and square centimeters in the denominator.

Algebraically it looks like this:

$$100,000\,\frac{\text{Newtons}}{\text{meter}^2} \times \left(\frac{1}{100}\,\frac{\text{meter}}{\text{cm}}\right)^2 = 10\,\frac{\text{Newtons}}{\text{cm}^2}$$

The factor equivalent to 1.0 is $1/100\,m/cm$. This factor needs to be squared (it's still 1.0) to get the units right and the number $1/100$ gets squared right along with the units. Therefore, the number 100,000 gets divided by 10,000, not by 100.

5.3: $14.7\,\text{pounds/in.}^2 \times 4.45\,\text{N/pound} \times [1./0.000645]\,\text{in.}^2/m^2 = 101,419\,N/m^2$

5.4: According to Eq. (5.1) the speed of sound at $-18\,°C$ becomes

$$v = 331.7 + 0.6(-18) = 331.7 - 10.8 = 320.9\,m/s,$$

and at $40\,°C$

$$v = 331.7 + 0.6(40) = 331.7 + 24 = 355.7\,m/s.$$

5.5 (a): Start with Eq. (5.1) and substitute Eq. (5.6) for T_C, the centigrade temperature.

$$v = 331.7 + 0.6[\frac{5}{9}(T_F - 32)] = 331.7 + \frac{1}{3}T_F - 0.6 \times \frac{5}{9} \times (32) = 312 + \frac{1}{3}T_F$$

and this is Eq. (5.7).

5.5 (b): Equation (5.7) has two terms. To convert to feet per second, they both need to be multiplied by the number of feet per meter. That number is

$$39.37\,\text{in./m} \times \text{ft/12 in} = 3.28\text{ft/m}.$$

i.e. multiply by 3.28 to get Eq. (5.8).

5.6: (a) time = distance/speed = $10\,m/\,(344\,m/s)$
    time = $0.029\,s$ or $29\,ms$.

(b) With races decided by hundredths of a second (tens of milliseconds) a delay of $29\,ms$ is significant. That is why a starter pistol is no longer used to start sprints. Instead there is a signaling device in each lane behind the runner.

5.7: The light can be considered to travel instantaneously. The delay of 2 s is all due to the relatively slow speed of sound. Therefore, distance = speed × time = 344 m/s × 2 s, or distance = 688 m away. That's 0.688 km, i.e., about 0.7 km which is about 0.4 miles.

5.8: 1128 ft/s × (1/5280) miles/ft × (3600) s/h = 769 miles/h. Notice how the units cancel numerator and denominator.

5.9: The difference in audible frequencies is 20,000 – 20, which is about 20,000 Hz. The difference in visible frequencies is $7 \times 10^{14} - 4 \times 10^{14}$ which is about 3×10^{14} Hz. Obviously the range of visible light is enormously wider in terms of differences. But when it comes to the behaviors of physical and biological systems, ratios are more important than differences. In terms of ratios, the range of audible sound frequencies is far greater than the range of visible light frequencies. The human ear needs to be able to deal with a low frequency like 20 Hz and with a high frequency like 20,000 Hz, i.e. 1,000 times greater. The human eye needs to be able to deal with red light, 4×10^{14} Hz, and with violet light, 7×10^{14} Hz, less than a factor of two greater. In other words, the range of visible light is less than an octave, but the range of audible sound is ten octaves, i.e. $2^{10} = 1024 \approx 1000$.

5.10: At room temperature for 20 Hz, wavelength = 344/20 = 17 m. That is more than 55 ft.

At room temperature for 20,000 Hz, wavelength = 344/20,000 = 17 mm. That is less than 3/4 in.

5.11: At room temperature, frequency = 344 (m/s)/1 (m) = 344 Hz.

5.12: The formula $\lambda = v/f$ conforms to the convention that independent variables are on the right side of the equals sign and the dependent variable is on the left. The speed of sound is determined by the environment, and the frequency is independent of the environment. The wavelength is what one gets depending on the other two variables.

5.13: The period is half a second; the fundamental frequency is 2 Hz. Successive solid circles may be the pressure peaks of each successive clap. The distance between the circles is the wavelength, $\lambda = vT$, or $344 \times 1/2 = 172$ m.

5.14: The unit "gpf" stands for "gallons per flush." To determine the water usage in a day you need to know the number N of "flushes per day." You do the calculation:

1.6 gallons/flush × N flushes/day = 1.6 N gallons/day.

Notice that the units of "flushes" cancels between numerator and denominator, giving you a final answer in units of "gallons per day."

5.15: The low-frequency parts of a complex sound would arrive at you sooner than the high-frequency parts. If the sound source is close to you it might not matter much, but the relative delay would be longer for a distant source. A distant explosion would sound to you like a whooosh with a pitch that rises. A voice in the distance would become unintelligible because the high and low frequencies would not be synchronized.

5.16: Transverse waves cannot propagate in water. To carry a transverse wave, a medium must have some resistance to being bent. Water has a lot of resistance to being compressed (longitudinal waves) but no resistance to being bent.

Chapter 6

6.1: According to Eq. (6.3) the requirement for cancellation is that the difference in distance be an odd number of half wavelengths. For this exercise, the difference in distance is 0.4 m. Therefore, for parts (a) and (b):

A distance of 0.4 m is one half of a wave length if the wavelength is 0.8 m. Then the frequency is 344/0.8 or 430 Hz.

A distance of 0.4 m is three halves of a wave length if the wavelength is 0.2667 m. Then the frequency is 344/0.2667 or 1290. Hz. This frequency is just 3 times 430. Hz.

(c) In theory, there is no limit to the number of magic wavelengths and frequencies. Any odd number of half wavelengths will work: 5/2, 7/2, 9/2, etc.

6.2: (a) There are two ways to do this exercise. Both give the same answer.

First way—use the period. The period is 1 ms and half a period is 0.5 ms. For cancellation the delay ought to be half a period. One gets a delay of 0.5 ms with a distance difference of 0.5×344 or 172 mm which is 0.172 m. Therefore, cancellation occurs if you put the second source $1.000 + 0.172 = 1.172$ m away from you, or else $1.000 - 0.172 = 0.828$ m away from you.

Second way—use the distance. The difference in distance ought to be half a wavelength. For a 1,000 Hz tone the wavelength is 344/1000=344 mm. Half a wavelength is 172 mm which is 0.172 m. Therefore, cancellation occurs if you put the second source $1.000 + 0.172 = 1.172$ m away from you or $1.000 - 0.172 = 0.828$ m away from you.

(b) In order to get complete cancellation the amplitudes of the two tones, as measured independently at the position of the receiver (namely at your position), need to be the same.

6.3: At low frequencies the wavelengths are larger than the dimensions of your room. If they are cancelled by a phasing error they are cancelled everywhere in your room. Getting a good bass response is a challenge for most audio systems. Recorded music is mixed with the left and right channels in phase for low frequencies to help with this situation. One thing that you really don't want to do is to cancel the bass.

When the wavelength is longer than the dimensions of the room, you can think of the loudspeakers as a pump, alternately compressing and expanding the entire volume of air in the room. For effective compression and expansion, the two speakers need to work together and not against one another.

By contrast, high frequencies have wavelengths that are much shorter than the dimensions of your room. High frequency sounds may cancel in one spot, but they reappear at another spot. This is a normal effect for any sound in any room; it is not special to stereophonic reproduction.

6.4: We know that the difference of the two frequencies is 5 Hz. If one of the frequencies is 440 then the other frequency must either be 445 or 435 Hz.

6.5: The beats between two sine waves are strongest if the amplitudes are equal because only then can complete cancellation occur. If the amplitudes are slightly different, there will always be a little bit of the larger wave left over when the phase relationship between the two waves leads to the greatest cancellation.

6.6: $\lambda = c/f$ or $3 \times 10^{8}/(6 \times 10^{14}) = 0.5 \times 10^{-6} =$ half a micrometer. A mirror needs to be polished so that bumps on its surface are smaller than that.

6.7: The wavelength of a 100-Hz wave is much longer than a few inches, but the wavelength of a 10,000-Hz wave is smaller than a few inches. The stone wall appears to be smooth to the 100-Hz wave, but it appears to be rough to the 10,000 Hz wave. The 10,000-Hz wave is diffusely reflected.

6.8: According to Fig. 6.8, sound waves bend toward the direction of cool air. Therefore, a channel is made by a layer of cool air between layers of warm air.

6.9: We use our ears to warn us of danger to ourselves or others. Apart from that, norml social interaction depends on verbal communication.

6.10: According to Fig. 6.11a, the slope of the input/output line is negative. The more the input pressure is positive, the more the output voltage is negative—and vice versa. The transducers described by Figs. 4.2 and 6.11 are actually the same except that Fig. 6.11 is *inverted*. It could be that the two transducers are identical except that the positive and negative wires coming from the transducer are reversed in Fig. 6.11.

Chapter 7

7.1: For the sixth mode there are six half wavelengths along the string, and five nodes.

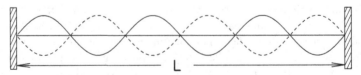

For Exercise 7.1. The sixth mode of vibration of a stretched string

7.2: $f = v_s/(2L)$ or 154 (m/s)/(1.4 m) = 110 Hz. The third harmonic has a frequency three times larger, 330 Hz.

7.3: The linear density (μ) is determined by the choice of guitar string. The tension (F) is determined by stretching the string on tuning the instrument. The length (L) is determined by the player who presses the string down on the frets.

7.5: The frequency is inversely proportional to the string length. We set up a proportion for the unknown length L.

$$82/110 = L/70.$$

The solution is $L = 52.2$ cm.

7.6: The top E frequency is four times greater or about 328 Hz. To get this high frequency the speed of sound needs to be four times faster.

7.7: Bend notes by stretching strings, increasing the tension. Some rock guitars have hinged bars (whammy bars) that stretch all the strings at once.

7.8: Double the tension and the speed of sound increases by a factor of the square root of 2. Therefore, the frequency increases by the same factor and becomes 141 Hz.

7.9: Start with F/μ. It is $\frac{\text{kg·m}/[(s)(s)]}{\text{kg/m}}$. The units of kg divide out and the denominator just becomes 1/m. Thus F/μ has units of m^2/s^2, and the square root of this is m/s as advertised.

7.10: The role of the guitar body is to radiate the sound. The frequencies of the sound are almost entirely determined by the string only. However, because the bridge and nut of the guitar body vibrate a little, the boundary conditions change slightly, and there is a tiny effect of the body on the playing frequencies. In this book, we always neglect that tiny effect.

7.11: In a wind instrument, the air pressure within the body of the instrument vibrates. In a drum, the drumhead (a membrane) vibrates. Both have modes of vibration. These modes are equivalent to standing waves. For instance, these modes have nodes at certain places.

7.12:

$$f_n = \frac{n}{2L}\sqrt{\frac{F}{\mu}}$$

7.13: The easiest solution is to recognize that the flat lines at 4 and 12 ms are separated in time by half a period. Therefore, half a period is $12 - 4 = 8$ ms and a full period is 16 ms. Therefore, the frequency is $1000/6 = 62.5$ Hz.

Chapter 8

8.1: (a) For open-open pipe $f = v/(2L) = 1130/(2) = 565\,\text{Hz}$.

(b) For open-closed pipe $f = v/(4L) = 1130/(4) = 282\,\text{Hz}$.

(c) The open-closed pipe has natural mode frequencies that are odd-integer multiples of the lowest mode frequency. Only modes with these frequencies are allowed to exist in this pipe. All the even numbered harmonics are missed.

8.2: (a) For open-closed pipe number of nodes = mode number -1

(b) For open-open pipe number of nodes = mode number -1,
    i.e., the relationships are the same.

8.3 Mode number four for a open-closed pipe has three nodes as shown $f = v/(4L) \times 7$:

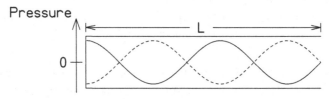

For Exercise 8.3. The fourth mode of air vibration in an open-closed pipe

8.4: The drawing should look like Fig. 8.2 except that there will be three nodes instead of one.

8.5: To put this matter in a clear light, we write an equation for the lowest frequency of a pipe that is open at both ends. Putting in the end correction explicitly, we find $f = v/[2(L + 0.61d)]$, where L is the measured length of the pipe and d is the diameter. The neat way to write this is to rearrange the denominator slightly to get.

$$f = v/[2L(1 + 0.61d/L)].$$

Now the formula looks just like our old friend, $f = v/2L$, with a denominator that is multiplied by something that is pretty close to 1.0 if d/L is small. This makes it clear that it is the diameter compared to the measured length, d/L, that causes the end-correction deviations.

8.6: The open-closed pipes need to be slightly longer because only one end correction is applied to them and in order to get a precise octave, the *effective* lengths of the two pipes need to be the same.

8.7: (a) The equation $f = v/(2L)$ can be solved for length L, to give $L = v/(2f)$. Therefore, $L = 344/(2 \cdot 16) = 10.75$ m. (b) One meter is 3.28 ft, and so $L = 35.3$ ft. (c) You can use an organ pipe that is half as long if you close off one end. Then, for an open-closed pipe $L = v/(4f)$. A 17- or 18-ft pipe will fit comfortably in the available space. (d) The solution does not come without cost. Closing off one end eliminates the even-numbered harmonics in a cylindrical pipe.

8.8: If an organ pipe is blocked off anywhere along its length, the pipe becomes an open-closed pipe with an acoustical length equal to the unblocked part. Therefore, if a designer wants a pipe to be 4 m long, but the pipe needs to have an acoustical length of only 1 m, a rigid barrier can be put into pipe at the 1-m mark and the remaining 3 m is just for show.

8.9: The tension and the mass density of a stretched string determine the speed of sound on the string. By contrast, the speed of sound of air in a pipe is determined by the air and its temperature. With air, we normally have to take what we get. To make a pipe more interesting, we could try to control the speed of sound in air. For instance, with refrigeration coils and a some blow torches we might be able to play a few different notes on the same organ pipe. Upon consideration, it does not seem like a very good idea.

Alternatively we might use a gas other than normal air. The speed of sound in helium is about three times the speed of sound in ordinary air. The speed of sound in nitrous oxide N_2O is about 80 % of the speed of sound in air.

8.10: After the initial impact, the air in the pipe continues to vibrate for a little while to make the tonal sound. If the pipe is pressed into your palm it becomes an open–closed pipe, but if the pipe was excited by extracting your finger, it is still an open–open pipe. The latter leads to a pitch that is an octave higher than the former.

8.11: $f = v/(4L) = 34400/(4 \cdot 60) = 143$ Hz, close to $D_3 = 147$ Hz.

8.12: The diameter converted to cm is $1.5 \times 2.54 = 3.81$ cm. The end correction is $0.305 \times 3.81 = 1.16$ cm.
$f = v/4L_{effective}$.
If $f = 65.4$ Hz, then $L_{effective} = 34400/(4 \times 65.4) = 131.50$ cm.
Taking the end correction into account, we must cut a length of pipe 130.34 cm long.

8.13: (a) For the longest possible cylinder (287 mm), expressing the speed of sound in mm/s, $344000/(4 \cdot 287) = 299.7$ Hz. The shortest cylinder is 232 mm shorter, or 55 mm long. Then, $344000/(4 \cdot 55) = 1564$ Hz. (b) The frequency ratio is 1564/299.7 or 5.2. This ratio is greater than the ratio for two octaves, which would be a ratio of 4.

Chapter 9

9.1: The problem only defines the relative amplitudes. It does not specify the absolute amplitude of any harmonic. Therefore, we can arbitrarily choose the amplitude of the fundamental to be 1.0.

Then

$$\text{Signal} = 1.0 \sin(360 \cdot 250\,t + 90) + 0.5 \sin(360 \cdot 500\,t + 90) +$$
$$0.333 \sin(360 \cdot 750\,t + 90) + 0.25 \sin(360 \cdot 1000t + 90).$$

9.2: The easy way to think about the change is to realize that the original complex signal looks like the fundamental except that the third harmonic puts dimples in each positive and negative peak. When the amplitude of the third harmonic grows to be as large as the fundamental, the dimples become large enough to cancel the fundamental entirely and bring the waveform to zero momentarily, as shown in the figure.

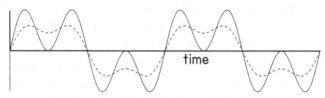

For Exercise 9.2. The *dashed line* is the reference. The *solid line* is the answer

9.3: (a) After 1/200 s, the 200-Hz wave has executed one complete cycle and is ready to begin again.

After 1/200 s, the 600-Hz wave has executed three complete cycles and is ready to begin again.

Therefore, after 1/200 s the sum of those two waves is ready to begin again and it has executed one complete cycle. Thus the period is 1/200 s. These facts are unrelated to the relative phases of the two components.

(b) A wave with components at 200 and 500 Hz has a period of 1/100 s.

(c) A wave with components at 200 and 601 Hz has a period of 1 s.

(d) A wave with components at 150, 450, and 750 has a period of 1/150 s. The wave consists of the first three odd-numbered harmonics only.

(e) A wave with components at 220, 330, 550, and 660 Hz has a period of 1/110 s. The fundamental at 110 Hz is missing and the periodicity comes from harmonics 2, 3, 5, and 6.

9.4: There is no perfect way to identify a psychological sensation like pitch with a physical measurement like frequency such that pitch can be expressed in Hz. What is done in practice is to define pitch in Hz as the frequency of a sine tone that leads to the same sense of pitch, i.e., the same height on a musical scale. For instance, we might play a 400-Hz sawtooth to a listener and ask the listener to adjust the frequency of a sine tone so as to match the pitch of the sawtooth. Suppose that the listener adjusts the sine frequency to be 401 Hz. Then we say that for this listener, on this experimental trial, the pitch of the 400-Hz sawtooth is 401 Hz.

9.5: Both the amplitude spectrum and the phase spectrum contribute to the shape of the waveform. Ohm's law of phase insensitivity applies to the *sound* of the waveform, not to its shape on an oscilloscope.

9.6: The next three terms are:

$$\frac{1}{4} \sin(360 \cdot 1600\, t + 180) + \frac{1}{5} \sin(360 \cdot 2000\, t + 180) + \frac{1}{6} \sin(360 \cdot 2400\, t + 180).$$

9.7 Drill answers:

(a) Four harmonics; (b) four lines on a spectrum analyzer; (c) second harmonic has the highest amplitude; (d) 5 units; (e) 60°; (f) Yes, a phase can be negative; (g) Yes, 240° is equivalent to −120° because those two angles differ by 360°; (h) The period is 1/200 s; (i) the quantity ft must have physical units of cycles. Therefore, if f is in Hertz, then t must be in seconds.

Chapter 10

10.1: (a) The square of $(1/3)$ is $1/9$.

(b) $10 \log(1/9) = -10 \log(9) = -10 \times (0.954) = -9.54\,\text{dB}$

10.2(a): From Eqs. (10.1) and (10.2) we know that $I \propto A^2$ and that $I \propto 1/d^2$. Substituting the first proportionality into the second, it follows that $A^2 \propto 1/d^2$, and the square root leads to $A \propto 1/d$, as advertised.

10.3(a): $I = P/(4\pi d^2) = 10/(4\pi 10^6) = 8 \times 10^{-7}\,\text{W/m}^2$
(b): $10 \log(8 \times 10^{-7}/10^{-12}) = 59\,\text{dB}$.
(c): An airplane in flight is perhaps the only condition where one finds 1 km of empty air in all directions, and an airplane can make 10 W of acoustical power. Alternatively—maybe a skydiving brass band?

10.4: Intensity dependence on distance (d) from the source for worlds with different dimensionality (D). As dimensionality goes from 3-D to 2-D to 1-D, the wavefront spreading law goes from (inverse second power, $1/d^2$) to (inverse first power, $1/d$) to (inverse zeroth power, 1). Zeroth power means no dependence on distance at all. Therefore, in a one-dimensional world there is no intensity loss due to wavefront spreading. The only losses are due to friction (absorption) in the medium of transmission.

10.5: $10 \log(30/1) = 10 \times (1.477) = 14.8 \, \text{dB}$.

10.6: $L_2 - L_1 = 10 \log(I_2/I_1) = 10 \log(2) = 10 \times 0.301 = 3 \, \text{dB}$.

10.7: (a) The statement of the situation in the exercise is that $20 = 10 \log(I_{\text{tmb}}/I_{\text{flute}})$
Therefore, $I_{\text{tmb}}/I_{\text{flute}} = 10^2 = 100$

(b) $I_{\text{tmb}}/I_{\text{flute}} = [A_{\text{tmb}}/A_{\text{flute}}]^2$, and so $[A_{\text{tmb}}/A_{\text{flute}}] = 10$.

10.8: We begin with the statement that

$$L_2 - L_1 = 10 \log(I_2/I_1)$$

Now $(I_2/I_1) = (A_2/A_1)^2$, so $L_2 - L_1 = 10 \log[(A_2/A_1)^2]$.
But the log of a square is twice the log, and so $L_2 - L_1 = 20 \log(A_2/A_1)$

10.9:

$$130 = 10 \log(I/10^{-12})$$
$$13 = \log(I/10^{-12})$$
$$I = 10^{13} \times 10^{-12} = 10 \, \text{W/m}^2.$$

Chapter 11

11.1: (a) The ear canal is said to be about 2.5 cm long. It resembles an open-closed pipe. Therefore, the resonant frequency is

$$f = v/(4L) = 34400/(4 \times 2.5) = 34400/10 = 3440 \, \text{Hz}.$$

(b) The second mode of such a pipe has the frequency

$$f = 3v/(4L) = 3 \times 34400/(4 \times 2.5) = 10,320 \, \text{Hz}.$$

11.2: The speed of sound in air is 344,000 mm/s. Five times that speed is 1,720,000 mm/s. Therefore, the time to go twice the distance of 35 or 70 mm in the fluid is

$$\text{time} = 70/1,720,000 = 0.000,041 \, \text{s} = 41 \, \mu\text{s}$$

11.3: The acoustic reflex is triggered by a whistle blowing next to your ear and other continuous sounds containing frequencies in the range from 1,000 to 4,000 Hz.

11.4: (a) For 440 Hz, about 9 mm from the apex. (b) For 880 Hz, about 13 mm from the apex.

11.5: Expect intervals of 1/440 s, 2/440 s, 3/440 s, and so on.

11.6: The lost hair cells must be those near the oval window.

11.7: There will be more interference when an intense 1,000-Hz tone masks a weaker 1,300-Hz tone because excitation patterns on the basilar membrane are unsymmetrical with greater excitation extending toward the base, i.e., toward the direction of higher frequency places.

11.8: For 5 mm,

$$f_c = 165(10^{(0.06)\cdot(5)} - 1) = 165(1.995 - 1) = 164\,\text{Hz}.$$

For 10 mm,

$$f_c = 165(10^{(0.06)\cdot(10)} - 1) = 165(3.981 - 1) = 492\,\text{Hz}.$$

etc.

11.9 (a): We start with the Greenwood equation,

$$f_C = 165(10^{az} - 1)$$

Our goal is to get z by itself on one side of the equation. We proceed systematically, doing what we can to move in the right direction at each step.

$$f_C/165 = 10^{az} - 1$$
$$1 + f_C/165 = 10^{az}$$
$$\log_{10}(1 + f_C/165) = az$$
$$\tfrac{1}{a}\log_{10}(1 + f_C/165) = z$$

Done!

(b): If $f_C = 440$ then, with $a = 0.06$,

$$\frac{1}{0.06}\log_{10}(1 + 440/165) = z = 9.4\,\text{mm}.$$

11.10: In the normal human auditory system there are half a dozen stages of neural processing between the cochlea and the final destination in the auditory cortex. Every one of these stages is tuned in frequency, with slightly different spatial regions for tones of different frequency. To best simulate a normal system, a cochlear nucleus implant ought to have multiple electrodes that the surgeon can place in appropriate regions of the midbrain. Cochlear nucleus implants are, in fact, made in this way.

11.11: (a) Microphone. (b) Microphone, Speech processor, Transmitter. (c) About one turn.

Chapter 12

12.1: Horizontal axis is frequency in Hz; vertical axis is level in dB.

12.2: About 41 phons.

12.3: The tone is inaudible. The point at 200 Hz and 10 dB is below the threshold of hearing curve.

12.4: The threshold of hearing is 0 phons whatever the frequency. That is the definition of zero phons. A sound level of 0 dB is also related to the threshold of hearing by definition, but the dashed line in Fig. 12.1 makes it clear that the actual decibel level for threshold depends on frequency. So long as the frequency is not too low or high, the threshold (dashed line) is rather close to 0 dB, i.e. close to 10^{-12} W/m^2.

12.5: From the equal-loudness contours, about 32 dB.

12.6: Choose 6,000 Hz. The equal-loudness contours dip down near 6,000 Hz, suggesting that this is a loud tone by itself. That is not true for 100 Hz. Also, this tone will excite different neurons from the 1,000-Hz tone. That cannot be said for the 1,100-Hz tone. Because neurons are compressive, you would prefer to spend your energy on different neurons.

12.7: (a) The equation says:

$$\psi \propto I^{0.3} \qquad\qquad (12.1)$$

Therefore,

$$\psi_2/\psi_1 = (I_2/I_1)^{0.3}$$

We are told that the ratio of the intensities is 10 to 1. Therefore,

$$\psi_2/\psi_1 = 10^{0.3} = 1.995 \approx 2.$$

(b)

$$\psi_2/\psi_1 = (I_2/I_1)^{0.3} = 50^{0.3} = 3.2$$

Therefore, 3.2 times louder.

12.8: (a) We know that the intensity ratio raised to the 0.3 power is 4. Therefore, the intensity ratio is 4 raised to the power 1/0.3, which is 102. Therefore, about 100 times. (Because 4 is the square of 2, we might have expected to find that two doublings would require multiplying the intensity by 10 twice.)

(b) $10 \times 10 \times 10 \times 10 = 10,000$. Therefore, 10,000 times.

12.9: The dip in equal loudness contours is caused by the resonance of the ear canal near 3,500 Hz.

12.10: From Eq. (12.7) we calculate that a level difference of 7 dB corresponds to a loudness ratio of 1.6. Thus, the trombone sound is 1.6 times louder.

12.11: In order to understand the effect of road noise on the loudness of a car radio, we need to assume that the auditory system can somehow separate the sound of the radio from the background noise and evaluate the loudness of the radio independent the noise. Just considering the total neural firing rate can never explain the effect. It is very likely that separating the sound of the car radio depends on the synchrony of neurons as they respond to the radio. The road noise suppresses the neural response that is synchronized with tones from the radio. When the road noise is gone the neural response is not suppressed.

12.12: Loudness is defined as a psychological quantity, in contrast to intensity, which is a physical quantity. Therefore, if it sounds louder, it really is louder, regardless of how the intensity might or might not change.

12.13: The noisy fan is 8 times louder than the quiet fan. 4 sones/ 0.5 sones = 8. The purpose of the sone scale is to make comparisons like this really easy.

12:14: In comparison with 1,000 Hz, the contours at 62 Hz are very close together. Therefore, the loudness increases rapidly as the sound level increases. At 62 Hz, as the sound level increases from 50 to 60 dB, the loudness level increases from 20 phons to 40 phons—twice as large an increase.

Chapter 13

13.1: 200 Hz appears at 5 mm from the apex and 2,000 Hz at 17 mm from the apex. The maxima for these two tones are 12 mm apart. That is a huge distance.

13.2: For Part a the histogram shows the following numbers of interspike intervals:
For 1 cycle (1/200 s) 5 intervals
For 2 cycles (2/200 s) 4 intervals
For 3 cycles (3/200 s) 2 intervals
For Part b, the histogram is
For 1 cycle (1/200 s) 2 intervals
For 2 cycles (2/200 s) 1 interval
For 3 cycles (3/200 s) 1 interval

For 4 cycles (2/200 s) 1 interval
For 5 cycles (3/200 s) 1 interval

13.3: These components look like harmonics 5, 6, and 7 of 150 Hz. The integers 5, 6, and 7 are consecutive and not very large. The pitch is very likely to be 150 Hz.

13.4: (a) This tone looks rather like harmonics 2, 3, and 4 of 200 Hz, but the harmonics are mistuned. We get a decent fitting template if we say that the template has a fundamental frequency of 620/3 or 206.7 Hz. The template expects a second harmonic at 2×206.7 or 413 Hz, and this is flat compared to the real component at 420 Hz. The template expects a fourth harmonic at 4×206.7 or 826.6 Hz, and this is sharp with respect to the real component at 820. In the end, the template is too low for the second harmonic, too high for the fourth harmonic, and just right for the third. The brain regards this template as a reasonable compromise and perceives this tone to have a pitch of 206.7 Hz.

(b) This problem is an extension of Exercise 13.4 (a). More mistuned harmonics (five and six) are added to the top end. If we think about a template with a fundamental frequency of 206.7 Hz, the sixth harmonic has a frequency of 1,240 Hz, and this is too far above the real component at 1,220 Hz. We are going to have to decrease the fundamental frequency of the template to get a better fit. We fit the center component (looking like the fourth harmonic) by choosing a fundamental frequency of $820/4 = 205$ Hz.

13.5: An analytic listener recognizes that the only difference between tones "1" and "2" is that 800 Hz component goes down to 750 Hz. This listener says that in the sequence "1" followed by "2" the pitch goes down. Most listeners hear Experiments a and b in this way.

A synthetic listener interprets tone "1" as the 4th and 5th harmonics of 200 Hz. This listener interprets tone "2" as the 3rd and 4th harmonics of 250 Hz. This listener says that in the sequence "1" followed by "2" the pitch goes up. The continuity in Experiment b should cause the synthetic listener hear the sequence differently. It should help him hear analytically if he wants to hear like most people.

The ambiguous sequence in Experiment a appears as demonstration number 25 on a compact disc entitled "Auditory Demonstrations," by A.J.M. Houtsma, T.D. Rossing, and W.W. Wagenaars, and is available from the Acoustical Society of America. It was reviewed by W.M. Hartmann, "Auditory demonstrations on compact disk for large N," J. Acoust. Soc. Am. *Review and Tutorial* **93**, 1–16 (1993).

Chapter 14

14.1: (a) $20 \log_{10}(1.18/1) = 1.44$. Thus, about 1.4 dB.

(b) $20 \log_{10}(0.68/0.5) = 2.67$. Thus, about 2.7 dB.

14.2: (a) The wavelength of a 100-Hz sound is about 3.4 m. This is about 20 times the diameter of the head. The wavelength is so much larger than the head diameter that the sound diffracts around the head and is just as large in the far ear as it is in the near ear. (b) To use the interaural level difference at a low frequency like 100 Hz, it would be necessary to have a head about five times larger than normal.

14.3: From Fig. 14.2 it looks as though there is about 1.8 dB separating the 10° curve from the 45° curve at 2,000 Hz. If the smallest detectable difference is 0.5 dB, then the range from 10° to 45° is divided into about 1.8/0.5 or 3.6 intervals. On average these intervals correspond to $(45 - 10)/3.6$ or 10°. Thus, the ILD is seen to provide only a crude estimation of azimuthal location.

14.4: (a) The left-hand side is time. The right-hand side is a distance divided by a speed, i.e., (meters) divided by (meters/second). The units of meters cancels in numerator and denominator leaving us with $1/(1/s)$, which is simply seconds.

(b) For 30°, 45°, and 60°, the ITD values from Eq. (14.1) are 382, 540, and 661 μs, respectively.

14.5: Localization is better in the horizontal plane in that it is possible to discriminate between two locations that are closer together in the horizontal plane. A listener can judge differences of about 1.5° in the forward direction in the horizontal plane, but can only judge differences of about 4° in the vertical plane.

14.6: The outer ear acts as a direction-dependent filter. As a result, a sound that comes from the back has a different spectrum, as it reaches the eardrum, compared to a sound that comes from the front.

14.7: The most important frequencies in front–back localization are above 6,000 Hz. Older people tend to lose high-frequency hearing, and many of them do not hear frequencies higher than 6,000 Hz. Because important speech frequencies are lower than 6,000 Hz, you would expect older listeners to lose front–back discrimination while maintaining good understanding of speech. Left–right (azimuthal) localization does not depend on such high frequencies. Low frequencies near 500 Hz tend to dominate.

14.8: According to the precedence effect, the first arriving sound determines the perceived location of the sound, even if later arriving sounds—early reflections and reverberation from a room—seem to point to different locations.

Chapter 15

15.1: Sound is transmitted through the wall by means of the studs. The drywall on the noisy side shakes the studs, and the studs shake the drywall on the other side. Putting fiberglass insulation in the gaps between the studs does not change this sound transmission process at all and is not an effective noise reduction method. What is effective is to use a staggered-stud wall where no stud is in physical contact with both sides of the wall. Once there is no contact between the wall surfaces it makes sense to add the fiberglass absorber.

15.2: Many rooms are acoustically bad because the heating, ventilating, air conditioning (HVAC) systems are poorly installed and make a lot of background noise. Other rooms may suffer from too much reverberation. Many environments have public address systems that are incorrectly adjusted or incorrectly used.

15.3: Compared to a plaster wall, a brick wall is not rough on an acoustical scale. The small bumps and valleys of a brick wall are not comparable to the wavelength of audible sound, and they do not produce diffuse reflections.

15.4: Decibels again—just when you thought it was safe to forget about them. If 99 % of the intensity is absorbed then 1 % of the intensity remains. The attenuation is found from the ratio of what remains to the initial amount. That ratio is 1 % or 0.01.

$$L_{remains} - L_{initial} = 10\log(0.01) = -20.$$

15.5: Nuclear reactions boil water to make steam to turn a turbine to run a generator to make electricity. Nothing about the final fate of all this energy is fundamentally changed by this generating process.

15.6: $T_{60} = 0.16\frac{V}{A_T} = 0.16\frac{2000}{218} = 1.5$ s.

15.7: To calculate the reverberation time using the Sabine equation we need the volume and total absorbing area. The volume is easy: 6.3 by 7.7 by 3.6 or 174.6 m^3.

The total absorbing area requires more work. There are two kinds of surfaces, plaster and vinyl on concrete. According to the table, the absorption coefficients at 500 Hz are 0.05 and 0.03, respectively.

There are two walls 6.3 by 3.6 m, each with an area of 22.7 m^2.
There are two walls 7.7 by 3.6 m, each with an area of 27.7 m^2.
There is a ceiling and a floor, 6.3 by 7.7 m, each with an area of 48.5 m^2.
Therefore,

$$A_T = 2 \times 22.7 \times 0.05 + 2 \times 27.7 \times 0.05 + 48.5 \times 0.05 + 48.5 \times 0.03 = 8.92\, m^2.$$

Then from the Sabine equation,

$$T_{60} = 0.16(174.6)/(8.92) = 3.1 \text{ s}.$$

That is a long reverberation time. In fact, the description of the room corresponds to a reverberation room used in acoustical testing.

15.8: The purpose of a shower is to provide an environment for singing, and most of the energy in the singing voice is below 500 Hz. We will assume that all surfaces are glazed tile and use an absorption coefficient of 0.01.

(a) Assume that the dimensions are 3 ft by 3 ft by 7 ft high. Assume that the shower has a glass door that also has an absorption coefficient of 0.01. Then the volume is

63 ft^3, and the total absorbing area (A_T) is $4 \times 3 \times 7 \times 0.01 + 2 \times 9 \times 0.01$. The A_T is thus 1.02 ft^2.

Sabine's formula involves $V/(A_T)$, which is 63/1.02 or 61.8 ft. To use the formula we need to convert that dimension to meters, and we find 18.8 m. Thus, $T_{60} = 0.16 \times 18.8 = 3$ s, and that makes a wonderfully long reverb time. Unfortunately, the human singer inside the shower adds absorption and reduces the reverberation time.

(b) Assume that everything is the same as in (a) except that there is no door. The shower is open on one side and that leads to an additional absorbing area of $3 \times 7 \times 1 = 21$ ft^2! Thus, $T_{60} = 0.16 \times 0.87 = 0.13$ s, and that is no fun at all.

15.9: If all the absorption coefficients are 1.0, then 100 % of the sound is absorbed on the walls. Such a room is an idealization and would be called "perfectly anechoic." Alternatively, such a room is really no room at all because it behaves just like empty space. With 100 % of the sound absorbed there is no sound reflected back into the room and the reverberation time is zero. But the Sabine formula says that $T_{60} = 0.16 \, V/(A_T)$, where A_T would be the total surface area in the room (multiplied by 1.0). All of the quantities are finite numbers, and the formula predicts a finite reverb time, contrary to the truth about this unusual, ideal room.

15.10: First, it is assumed that volume and total absorbing area are computed in the metric system. Then, the numerator (volume) has dimensions of cubic meters and the denominator (area) has dimensions of square meters, so the fraction has dimensions of meters. In terms of a dimensional equation:

$$\text{meters}^3/\text{meters}^2 = \text{meters}.$$

15.11: The volume of B is 8 times the volume of A. The total absorbing area of B is 4 times that of A. Consequently, the reverberation time of B is twice that of A.

Chapter 16

16.1: Charge is the basic quantity of electricity. Voltage is a force on a charge. The force of voltage causes charge to flow. Current is the rate of charge flow.

16.2: Both the dynamic microphone and the generator use the electromagnetic generator principle to convert motion into an electrical voltage.

16.3: No, not possible. Although a permanent magnet, like the magnet in a loudspeaker driver or a steel compass needle, has no *apparent* electrical current there actually is a microscopic current. On an atomic level the magnetism of a permanent magnet is caused by tiny electron currents and spins in the magnetic material.

16.4: For the loudspeaker to operate, the electrical current comes from the power amplifier and flows through the voice coil. The magnetic field comes from permanent magnets that surround the voice coil. The force is transmitted to the cone or dome of the speaker, having large surface areas, in order to move air creating an acoustical pressure wave.

16.5: The electron beam in the CRT is a flow of electrons, and that is a current. The loudspeaker has to have permanent magnets in order to work at all, and some of the magnetic field can leak out and cause a force on the beam in a nearby CRT. This is the motor principle, even though the current is not actually carried by a wire. The force bends the electron beam, which causes objects to appear in slightly wrong places on the screen, thus distorting the picture. Because of this effect, loudspeakers intended to go next to computer monitors are shielded to reduce the leakage of magnetic field.

16.6: The coil is wrapped around the magnet and there would seem to be no relative motion between the wire of the coil and the magnet. Therefore, it would seem that the generator principle would not apply. However, we know that these pickups *do* work, and so we have to look further. We begin by noting that the generator principle only requires motion between the coil of wire and the magnetic *field*. That's the key. What actually happens is that the motion of the steel guitar string close to the magnet of the pickup causes a small change in the amount of magnetic field that threads its way through that magnet. Thus the magnetic field is changing, and the coil of wire is right there in the presence of this changing field. Thus a voltage is induced in the coil of wire.

16.7: The coil is lighter than the magnet. Attaching the coil to the diaphragm, instead of the magnet, would lead to less inertia and more relative motion between coil and magnetic field.

16.8: (a) Physically, both the motor and the generator consist of a coil of wire in a magnetic field. The coil of wire is attached to a shaft which transmits mechanical rotation. The wire has two ends, and an electrical current can flow in and out. If a current passes through the coil of wire, there is a force on the wire because of the motor principle. That force can turn the shaft. That makes a motor. It can do your laundry or start your car engine. In the reverse operation, an external source of mechanical energy turns the shaft, and a voltage is induced between the two ends of the wire because of the generator principle. (b) Yes, a loudspeaker can be used in reverse to serve as microphone. It's easy to show if you connect the two terminals of a loudspeaker driver (a woofer works very well) to the input to a sensitive oscilloscope. If you then talk into the loudspeaker you will see the generated voltage on the 'scope.

16.9: When the wire of the voice coil melts it no longer carries any current. With no current, the speaker cone feels no force and the loudspeaker driver makes no sound at all. The only practical remedy is to replace the driver, which could be expensive.

Chapter 17

17.1: Both 60 and 50 Hz are low frequencies where the equal loudness contours are rising steeply. For such low frequencies, the lower the frequency the harder it is to hear. Therefore, the answer to this exercise is, Yes, all other things being equal it is harder to hear hum at 50 Hz than 60 Hz. There is more. Hum normally includes second and third harmonics—possibly even stronger than the fundamental. Then, the fact that 100 Hz is harder to hear than 120 Hz, and 150 Hz is harder to hear than 180 Hz also would make the European hum less objectionable.

17.2: In order to avoid distortion, the output of a power amplifier must have the same shape as the input. When an amplifier is overdriven, the amplifier has reached the limit of its output voltage. It cannot produce a voltage that is high enough (or negative enough) to capture the peak (or valley) required by the shape of the input. The resulting change in waveform shape is nonlinear distortion.

17.3: (a) Chords consist of many frequencies—the fundamentals and harmonics of all the notes in the chord. Nonlinear distortion causes difference tones among all these components in the chord. That leads to a mass of spectral components that are the characteristic sound of the distorted rock guitar. (b) What sounds bad depends on the context. In heavy metal music distortion is used as an artistic device.

17.4: You should choose the 500-Hz tone (d). With the four sources available to you, you can only test for harmonic distortion. You will listen for spectral components in the output that do not occur in the input. With the 500-Hz tone you will easily hear the second and third harmonics at 1,000 and 1,500 Hz. The harmonics of the 5,000-Hz tone (c) are too high to hear easily. The complex tone (b) already has harmonics in the input. Therefore, if harmonics are heard in the output you learn nothing. The broadband noise (a) has components at all frequencies and so distorted noise sounds just like undistorted noise.

17.5: The fundamental frequency is 1,000 Hz.

17.6: Harmonic distortion products occur at 2,000, 3,000, 4,000, ... Hz and 2,400, 3,600, 4,800, ... Hz. Important difference tones are $f_2 - f_1 = 1200 - 1000 = 200$ Hz, $2f_1 - f_2 = 2000 - 1200 = 800$ Hz, and $3f_1 - 2f_2 = 3000 - 2400 = 600$ Hz. Important summation tones can be found by replacing the minus signs above with plus signs. Therefore, they include 2,200, 3,200, and 5,400 Hz. Physically, the summation tones have the same amplitudes as the corresponding difference tones, but the difference tones are much more audible. One reason for the greater prominence of the difference tones is that the difference tones have frequencies that are lower than the input sine tones and so they are not as efficiently masked by the input sine tones.

17.7: When frequency f_2 increases from 1,200 to 1,250 Hz, the difference tone $2f_1 - f_2$ decreases from 800 to 750 Hz. If this difference tone is prominent, it may even change a listener's perception from an upward pitch change to a downward change.

17.8: Assuming that the tolerable distortion is 0.1%, the dynamic range is $90 - 10 = 80$ dB.

17.9: Fundamentally, there is no difference between the 60-Hz electrical power from the outlet in the wall and a 60-Hz audio signal from a power amplifier. Both are called "alternating current," both have the same waveform (sine wave) and frequency. The 120-V power from the wall has an amplitude of $120\sqrt{2} = 177$ V. That is much higher the voltage used to drive loudspeakers in a home audio system, normally only a few volts. Therefore, a home audio system does not have enough power to run a toaster (about 1,000 W). On the other side, if you were to plug a standard loudspeaker into a wall outlet, you would be driving the loudspeaker at a rate of $(120)^2/8 = 1800$ W ... briefly ... because the loudspeaker coil would burn out very quickly.

17.10: (a) The tall peaks in the spectrum have frequencies that are integer multiples of 350 or 450 Hz. They occur at 700, 1,050 Hz, ... and 900, 1,350 Hz ... These are harmonics which tells you that there are two periodic waveforms here.
(b) It's a good bet that there are two physical horns, each one making a periodic wave.
(c) The frequency ratio is 450/350 or 1.286. Table D1 indicates that the closest musical interval is an equal-tempered major third with a ratio of 1.26.
(d) If we define tall peaks as peaks that are no smaller than 20 dB less than the -60 dB peak (i.e. 20 dB down) at 350 Hz (i.e. 20 dB down), then all the peaks are accounted for by assuming that they are harmonics.
(e) If we decide to include all the peaks higher than the 30-dB down point, then we see distortion components at frequencies (Hz) given by $2f_1 - f_2 = 250$, $f_1 + f_2 = 800$, $4_1 - f_2 = 950$, $3f_1 + 2f_2 = 1950$, $4f_1 + 2f_2 = 2300$, $3f_1 + 3f_2 = 2400$, $6f_1 + f_2 = 2550$. There are probably more difference tones too, but they are masked by low frequency noise—likely wind noise. Interestingly, all the smaller peaks can also be accounted for with combination tone formulas like those above but different integers, sometimes large integers.

Chapter 18

18.1: The amplifier consists of a source of voltage that may be positive or negative. This voltage initially comes from positive and negative power supplies, which are the most massive part of the amplifier. The positive and negative voltages from the supplies represent the largest magnitudes of voltage that the amplifier will ever be able to produce. What the output voltage actually is at any given time is determined by the input signal, which can have a voltage that is much smaller than the output voltage. The process of controlling a large output voltage with a small control signal is called amplification, and it is normally accomplished with transistor amplifiers or integrated circuits that contain transistors.

18.2: (a) $20\log(100) = 40$ dB. (b) $20\log(1000) = 60$ dB.

18.3: For an 11-band, octave-band equalizer, the center frequencies of the bands, starting at 20 Hz, would be 20, 40, 80, 160, 320, 640, 1,280, 2,560, 5,120, 10,240, 20,480 Hz.

18.4: A valve is a device that controls a flow. A water valve turns the water on and off and regulates its flow. It's natural to refer to a vacuum tube as a valve because it can control the flow of electrical current in the output circuit by means of a small input voltage.

18.5: (a) The car radio has a tuner, preamplifier, tone controls, and power amplifier. It may have a CD drive. (b) The boom box is the same as the car radio but includes loudspeakers too. (c) A desk top computer has a CD drive and a sound card that includes a preamplifier for microphone input. A desk top computer normally is connected to speakers that include their own power amplifier. (d) A lap top computer has everything on the list except for a tuner and tone controls. (e) A camcorder has a microphone, preamplifier, a tiny power amplifier, and a tiny speaker. Digital cameras with audio are the same.

18.6: There is a school of audio recording that advocates *tight mikeing* whereby a large number of microphones are placed all around the orchestra to make close-up recordings of individual instruments or groups of instruments. Dozens of individual recordings feed separate channels of a mixer where they can be adjusted to taste and mixed with the signals from other microphones that have been placed to capture the ambiance of the environment. One can easily use 40 microphones and 40 channels.

Chapter 19

19.2: Choose a resonance frequency for the enclosure that is somewhat less than 60 Hz. By boosting frequencies where the driver is weak, you can extend the bass response of the system to lower frequencies.

19.3:

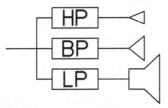

For Exercise 19.3. Three-way loudspeaker. The crossover has one input and three outputs From top to bottom, the drivers are tweeter, midrange, and woofer

19.4: Efficiency is output power divided by input power. For A: $1.5/20 = 7.5\%$. For B: $1/14 = 7.1\%$. Thus A is slightly more efficient, but for all practical purposes the two speakers have equivalent efficiency. [$10 \log(7.5/7.1) = 0.2$ dB—an inaudible difference.]

19.5: To calculate the output of the speaker in dB SPL we imagine that we put 1 W into the speaker. Then, from the description in Exercise 4, we expect to get 1/14 acoustical watt out of the speaker. To calculate the level of sound produced we imagine that this power, 1/14 W, is distributed over the surface of a sphere that is 1 m in radius. We further assume homogeneous distribution. The area of a sphere with radius R is given by $4\pi R^2$. For a 1-m sphere that is 12.6 m^2. A power of 1/14 W spread over a surface of 12.6 m^2 is an intensity of 5.68 mW/m^2. The threshold of hearing is 10^{-12} W/m^2 or 10^{-9} mW/m^2. Therefore, the level is

$$10 \log(I/I_0) = 10 \log(5.68/10^{-9}) = 97.5 \text{ dB SPL}$$

From a physical point of view this seems like a strange way to quote an efficiency, but it has obvious practical value to the speaker designer who wants to know how loud the speaker is going to sound.

19.6: Near the origin, where signal levels are low, the slope of part (b) appears to be about the same as the slope of part (a). Maybe (a) and (b) are really supposed to be the same device, but (b) is defective in some way.

19.7: At low frequencies the wavelength is long and the ratio of wavelength to driver cone diameter is large. That is the condition for good diffusion, and so you expect the frequency response for 45° to be similar to the response for 0°. Similar response at high frequencies would violate the rule for wave diffraction.

19.8: The plot shows that compared to the response at a mid frequency like 0.5 kHz, the response at the low frequency of 0.05 kHz has rolled off by about 6 dB.

Chapter 20

20.1: Obviously the two statements are equally accurate.

20.2: For a three-bit word, $N = 3$. From the formula we expect there to be eight possible values of the signal. These are:

000 001 010 011 100 101 110 111.

In a system with positive numbers only, these represent the numbers 0, 1, 2, 3, 4, 5, 6, and 7 respectively.

In the audio standard that uses negative numbers, these represent the numbers 0, 1, 2, 3, −4, −3, −2, and −1 respectively.

20.3: If there are 44,100 samples per second per channel and there are two channels in a stereo recording, then stereo requires 88,200 samples per second. If there are 16 bits in a sample, then a stereo recording requires 1,411,200 bits per second. If there are 8 bits in a byte, then you can find the number of bytes by dividing the number of bits by 8 or 176,400 bytes per second.

These values of the data rate apply to the compact disc format, which uses no data compression. Data compression techniques can greatly reduce the data rate and storage requirements, but information is lost in that process. In actuality, the compact disc format requires a higher data rate than we have calculated because this format includes redundant information that enables the playback system to recover from errors in reading the data. The error correction codes need to be added to the minimum set of data represented by 88,2000 samples per second.

20.4: If the sample rate is at least twice the highest frequency, then there must be at least two samples for every cycle.

20.5: The telephone uses only a single channel (no stereo telephone) and the sample rate needs to be twice the maximum frequency or 10,000 samples per second.

20.6: Digital electronic music synthesizers compute the musical waveform from a formula, as controlled by the musician. This music has no existence as an analog waveform until it is converted to an audio signal by a DAC.

20.7: Reasonable people can differ on this question, but an important argument for the analog vinyl recording is that the ET s will quickly recognize what it is and figure out how to play it back. By contrast, information in the compact disc format is highly encoded and is further complicated by error correction codes, redundancy checks, and added data outside the audio domain. Getting the information off a compact disc requires that the reader have knowledge of these arbitrary encoding conventions or somehow figure out what they are. If we want the ET s to be able to read the compact disc we had better hope that they are very clever.

20.8: Because loudspeakers and headphones receive an analog signal from a power amplifier, it is hard to see how the format of the original program material—digital or analog—could matter at all. "Digital ready" would seem inappropriate.

20.9: Unless the sampling of the signal is somehow synchronized with periodicity in the signal itself (an unlikely event) the errors made by quantization are randomly varying like noise. We normally reserve the term "distortion" to describe a systematic, time-independent deformation of a waveform which would be the same on each cycle of a periodic signal.

20.10: (a) $2^8 = 256$. Thus, 256 different values. (b) $20 \log(256) = 48$. Thus the signal to noise ratio is 48 dB. That is not nearly good enough for audio. Older technologies, such as vinyl disks and magnetic tape, have better signal to noise ratios than that 20.11 : $1 + 4 + 8 + 32 + 64 = 109$.

20.12: A byte for the number 85 is 0101 0101. It is $1 + 4 + 16 + 64$.

20.13: Commercial analog recordings of music were frequently copied, but each generation had more noise than the previous generation. This loss of quality in reproductions caused consumers to buy original recordings. Digital recordings do not suffer this sequential degradation, and that makes frequent copying more attractive. Nevertheless, sharing files among friends had little effect on the music industry. What really changed the game was the Internet which made it possible for millions of people to share the same digital music file. The battle between consumers and the industry raged throughout the first decade of the twenty-first century.

20.14(c): Digital technology cannot be acausal in the truest sense. However, if we make a continuous recording of a sound and are willing to delay the output of the processing by 10 ms, we can look ahead 10 ms in doing the processing.

Chapter 21

21.1: The wavelength is given by $\lambda = c/f$, where c is the speed of light. Therefore, for 1 MHz, (a) $\lambda = 3 \times 10^8/10^6 = 3 \times 10^2 = 300$ m. For a frequency of 10^{11} Hz, the wavelength is (b) $\lambda = 3 \times 10^8/10^{11} = 3 \times 10^{-3} = 0.003$ m.

21.2:

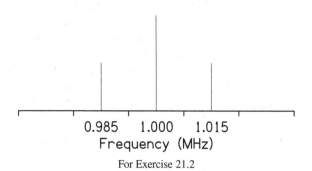

For Exercise 21.2

21.3: $6,000,000/30,000 = 200$ times wider.

21.4(c): An AM receiver is sensitive to the amplitude of the signal and when noise is added to the signal the amplitude includes that noise. A FM receiver is sensitive to the phase of the signal, particularly the rate of change of that phase. It is not sensitive to changes in the amplitude of the signal. When noise is added to the signal it apparently has less effect on the phase than it does on the amplitude.

21.5: *Back row of the opera:* The time required for a sound wave to go 150 ft can be found from the speed of sound, 1.130 ft/ms. The delay to the back row is therefore $150/1.130 = 133$ ms.
Broadcast: Because the circumference of the earth is about 36,000 km, any two points on the surface of the earth are separated by no more than 18,000 km. A radio wave goes that far in $18,000/(3 \times 10^5) = 60$ ms. Adding at most 20 ms for the

acoustical paths from singer to microphone and radio to you leads to no more than 80 ms. Therefore, you hear the singer before the audience in the back row.

21.6: To transmit audio or video information via a radio-frequency carrier, the carrier must be modulated. When the carrier is modulated it acquires sidebands. Although a carrier by itself is only a single frequency, transmitting the sidebands requires a frequency *region* or band. No other transmission can use that frequency region without causing interference. To avoid chaos, the use of frequency bands must be regulated. In the USA, for example, radio bands are assigned by the Federal Communications Commission, part of the Department of Commerce. The greater the rate of transmitting information, the faster the modulation needs to be. High modulation frequencies lead to large bandwidth. Because bandwidth is a scarce commodity, information transmission itself becomes scarce, and economic factors apply. This rule is completely general. It does not matter if one is transmitting audio by AM radio or if one is transmitting digital data by modulating an optical carrier (light beam). The rule is the same: Information transfer requires bandwidth.

21.7: A lightning bolt produces a brief intense burst of electromagnetic radiation. This leads to static in radio reception. Because it is a pulse, static has a broad spectrum. Thus, the power is spread over a wide frequency range. The tuned circuit at the front end of a radio receiver eliminates all of that broad spectrum except for the narrow band that happens to coincide with the desired signal. But since the signal is deliberately put in that band, whereas the static is only accidently in that band, the signal normally dominates the static. But not always—AM broadcasts are frequently interrupted by static.

21.8: Yes, there is agreement. For example, channels 7–13 must fit into a band with a width of $216 - 174 = 42\,\text{MHz}$. Because there are seven channels, 7, 8, 9, 10, 11, 12, and 13, they will fit if each of them is 42/7 or 6 MHz in width.

Chapter 22

22.2: Lung capacity is 1.32 gallons or 5.28 quarts.

22.3b: B and V are voiced plosives.

22.4: The nasals can be made with the lips tight shut, but a listener may not know which nasal is being pronounced. There is also a nasal aspiration, not on the list of phonemes, as in "hmmmm." Although these closed-lips phonemes would seem to have minimal content, a lot of information can be transmitted by pitch changes during phonation.

22.5: In theory, all vowels might sound the same if played backward, but speakers of English tend to make diphthongs out of all vowels, more or less depending on regional accents. By their nature, these diphthongs sound different if played backward. For standard American speech (e.g. network TV announcers) the EE and the OO tend to be the same backward and forward.

22.6: The great thing about answering questions like this one is that you don't have to know anything new. Just make the sounds and report what you do. The sequence goes from front to back.

22.8: There is no reason to change. The spoken vowel and whispered vowel both have the same formants as determined by the articulatory structures—mouth, lips, tongue. The difference between speaking aloud and whispering is only in the different excitation source from the lower parts of the vocal tract.

22.9: Low fundamental frequencies and high formants occur for cartoon characters in films and TV ads. For instance, there's the baby who smokes cigars.

22.10: The vowels have the most energy and appear darkest on the spectrogram. They have formants—alternating dark and light bands as a function of frequency. Formant transitions appear as dark bands that slide from a higher frequency range to a lower frequency range or vice versa. Fricatives appear as noise, broadband, or high frequency. They do not show formant structure—there is only one main frequency band. Plosives are like fricatives but they are brief and don't occupy much time on the horizontal axis.

Chapter 23

23.1: The trumpet player calls the notes "C,G,C,E,G,C"—the notes of a bugle call. Their frequencies are: 233, 349, 466, 587, 698, and 932 Hz.

(a) The notes called C correspond to frequencies: 233, 466, and 932. Each one is twice as high (factor of 2) as the previous one. A factor of two in frequency is an octave.

(b) From C_4 to G_4 is a ratio $349/233 = 1.498$. From C_5 to G_5 is a ratio $698/466 = 1.498$. These intervals are close enough to $3/2 = 1.500$ to be good intervals of the fifth.

23.2 (c): For $G4$: Resonances 3 and 6. (Resonance 9 is on the high side.) For $C5$: Resonances 4 and 8. For $E5$: Resonances 5 and 10. For $G5$: Resonance 6. (The second harmonic falls between resonances 11 and 12.) For $C6$: Resonance 8.

23.3: The resonances of a bugle are at successive integer multiples (2, 3, 4, ...) of a (nonexistent) base frequency. A cylindrical pipe that is open at both ends also has resonance frequencies that are successive integer multiples of a base frequency (Chap. 8). In that sense, the two systems are similar. However, the bugle is not a cylindrical pipe. It gets its harmonically related resonances from its bell.

23.4: If we represent a closed valve by the symbol 1 and an open valve by a symbol 0 then the possible positions can be represented by the answer to the first part of Exercise 20.2, namely:

$$000\ 001\ 010\ 011\ 100\ 101\ 110\ 111.$$

The missing configuration is 001, where only the third piston is depressed.

23.5: A tube with a flaring bell and mouthpiece is really quite different from a cylindrical pipe open at one end and closed at the other. The frequencies of the basic cylindrical pipe are substantially altered when the bell is added. Therefore, we cannot answer this question by appealing to the formula $(v/4L) \cdot (2n + 1)$ for the resonances of a cylindrical pipe.

A better answer appeals to the idea of scaling. Suppose that a trombone were absolutely identical to a trumpet except that every dimension of the trumpet is multiplied by 2 to make a trombone. Then every resonance of the trombone would be the same as the resonance of the trumpet except that the trombone resonance would occur for a wavelength that is twice as long. Doubling the wavelength reduces the frequency by an octave, and that would cause the trombone to sound an octave lower. In fact, the trombone is close enough to a scaled trumpet that this scaling argument is valid. It may strike you as surprising that the trombone is a scaled trumpet. The two instruments don't look the same. The trombone has a slide and the trumpet has valves. Acoustically, however, both are cylindrical tubes with flaring bells on the end. Whether extra cylindrical tube is added by extending a slide or by operating a valve is acoustically less important.

23.6: First, refer back to Exercise 23.2. The trumpet player and trombone player use the same resonances of the instrument. The instruments sound different because the resonances of the trombone are an octave lower than the trumpet. By contrast the French horn player uses higher-numbered resonances. Even though the instrument may be as long as a trombone, the playing frequencies are not low like a trombone. Using the higher-numbered resonances leads to notes of higher frequency.

23.7: Dents in a beat up bugle are small and don't change the basic shape of the instrument. They do change the details. The low frequency modes, which are most important in determining the playing frequency, have long wavelengths and they are not much affected by the dents. Dents can mistune the high-frequency modes and cause them not to line up well with harmonics of the tones. When high-frequency modes do not support the upper harmonics of the tone, a brass instrument sounds dull or stuffy.

23.8: Bozo is wrong as usual. The player does not voluntarily move his lips at a rapid rate. He uses his lip muscles to set a favorable lip configuration inside the mouthpiece. After that, his lips vibrate rapidly under the influence of a more or less steady stream of air. Vibration caused by a steady stream is fairly common. Venetian blinds will rattle in a steady breeze. The nozzle of a balloon may vibrate as air comes out of the balloon. You can blow a steady stream of air through your mouth and trill your tongue. The brass instrument player's lips do the same kind of thing, except their motions are organized by the influence of the horn. Of course, setting a configuration that favors lip vibration is not trivial. When a beginner first tries a brass instrument, he is likely to get no sound out of it at all—no lip vibration.

23.9: The resonance curves, such as those in Fig. 23.6, come from the linear response of the instrument itself . . . no musician involved. To measure this response you arrange to have a sinusoidal oscillating flow of air into the instrument so that the amplitude of the flow (measured in cubic centimeters per second) is constant as the frequency of the oscillations changes. Then you measure the response, which is the amplitude of the pressure oscillations in the mouthpiece. That response looks like Fig. 23.6, which can fairly be called a resonance curve because it shows a ratio of output to input as a function of input (driving) frequency. To use a more technical term, you can say that you have measured the magnitude of the "input impedance" of the horn as a function of frequency. Having found this linear response, we can now add the musician. We imagine a musician with lips of steel who can maintain a steady flow of air into the instrument through a fixed opening between his lips. Then the pressure in the mouthpiece would be proportional to the flow. But the real musician does not have lips of steel. Instead, the opening depends on the pressure in the mouthpiece. In the end, the resulting pressure in the mouthpiece depends on the resulting pressure in the mouthpiece, and that is a nonlinear process. The nonlinear process causes the lips to buzz at a frequency and with a waveform determined by the feedback from the horn with its tuned resonances.

Chapter 24

24.1: (a) For a tube like this, the resonances are given by the formula $(v/4L) \cdot (2n + 1)$. (Recall Chap. 8.) The odd-numbered resonances create odd-numbered harmonics in the tone. The clarinet actually differs from a cylinder because it has a bell, and because it has holes bored into its wall. Even if the holes are closed, they represent irregularities in the wall surface. Also the mouthpiece is tapered, and that is another irregularity. These irregularities are responsible for the even-numbered harmonics in the higher register of the clarinet. For the low register the wavelengths of low-numbered harmonics are long enough that the irregularities are hardly noticed. Then the ideal cylindrical pipe, with odd-numbered resonances only, is a better approximation to the real clarinet.

(b) If $v/(4L) = 147$, where $v = 344$ m/s, then $L = 344/(4 \cdot 147)$, and $L = 0.585$ m.

24.2: Consider the pressure waves for the first and second modes of a cylindrical pipe, open at one end and closed at the other, as shown in Chap. 8. One third of the way from the closed end the second mode has a node. If a hole is opened at this point in the pipe, nothing is changed so far as the second mode is concerned. The standing wave pattern for this mode has ordinary atmospheric pressure at that point whether the hole is there or not. Now consider the first mode. The first mode requires considerable pressure variation at the point that is one-third of the distance from the closed end. The first mode cannot survive with a hole at that point. Therefore, the register hole kills off the first mode while not affecting the second mode (frequency of three times the first).

24.3: For an open–open pipe the frequency of the fundamental is $v/2L$. For an open–closed pipe the frequency of the fundamental is $v/4L$. There's the octave.

24.4: From the study of Fourier analysis we know that waveforms that change abruptly in time have high-frequency components. Just imagine all the high frequencies it takes to synthesize a waveform that has a sharp corner. The abrupt closing of a double reed leads to a waveform with sharp corners. The relatively intense harmonics with high-harmonic number lead to a bright sound. Renaissance double reed instruments like the krumhorn make such an impulsive waveform that the high harmonics are very numerous and very intense. As a result such instruments are more than bright—they are buzzy.

24.5: (a) The free reeds and the vocal folds both vibrate at frequencies determined only by themselves, without any feedback from a resonator. The saxophone reed vibrates at a frequency that is largely determined by feedback from the horn. (b) In a double-reed instrument like the oboe, the two reeds vibrate against one another— like the two halves of the vocal folds.

24.6: For a pipe open at both ends the playing frequency should be $f = v/(2L)$. For the flute, $f = 344/1.2 = 287$ Hz. From Appendix E, you find that this frequency is between $C\sharp_4$ and D_4. In fact, the head joint—a tapered extension of the flute— adds some effective length, and the lowest note on the flute is actually C_4 with a frequency of 262 Hz.

24.8: Hints: Use $v = 34,400$ cm/s, and $V = (4/3)\pi r^3$, where r is the radius of the sphere. Use the approximation for the tone hole radius $a = r/2$, and remember that diameter D is twice r. Then turn the algebraic crank on Eq. (24.2) for the neckless Helmholtz resonator.

24.9: Remember what you learned in Chap. 13 about the pitch of a periodic tone with a missing fundamental.

24.10: (a) The fundamental appears to be about 262 Hz. (You can get good accuracy by finding the 10th harmonic and dividing its frequency by 10.) Appendix E shows that the closest note is C_4 (261.6 Hz), called "middle C." (b) One can see 19 harmonics out to 5,000 Hz. (c) The spectral components seem to be regularly spaced as expected for a harmonic spectrum, and so one expects the tone to be periodic (period $= 1/262$ s). (d) The second harmonic seems to be about 6 dB higher than the fundamental. For harmonica tones, the second harmonic is always strong, as expected for a free-reed instrument, but it is not always stronger than the fundamental. (e) The 12th harmonics seems to be 3 or 4 dB higher than the fundamental. That makes the harmonica a highly unusual instrument. For most instruments in their melody-range, harmonics as high as the 12th are much weaker than the fundamental. The strong high harmonics for the harmonica lead to the characteristic reedy sound.

Chapter 25

25.1: From the formula $f = v/(2L)$ we find that the speed of sound is $v = 2Lf = 2 \cdot 346 \cdot 440 = 304{,}480$ mm/s or about 304 m/s.

25.2: The lengths are 346, 326.6, 308.2, 290.9, 274.6, 259.2, 244.6, and 230.9 mm. The successive differences are 19.4, 18.4, 17.3, 16.3, 15.4, 14.6, and 13.7 mm. These differences are similar but not the same.

Note: The frequencies of the scale do not grow linearly—they grow geometrically with a common ratio. The frequencies of the string do not vary linearly with string length—they vary in a reciprocal way. Both these nonlinear functions produce greater frequency differences for greater frequencies.

25.3: For both the violin and the voice, there is a source and a resonator. The source vibration pattern and frequency are very little affected by the resonator. Source and resonator are largely independent in their operations. The role of the resonator is to filter and radiate the sound. The resonator has many peaks and valleys in its response function, which is responsible for the tone color. When a tone is played or sung with vibrato (periodic frequency modulation), the resonances lead to a frequency-modulation-induced amplitude modulation of the harmonics. This relationship between source and resonator is very different from a wind instrument, either brass or woodwind, where the resonator provides feedback that helps determine the vibrating frequency of the source.

25.4: Your graph should show about five or six cycles of a sine wave with an amplitude of about 12 Hz, hence extending from 380 to 404 Hz.

25.5: The acoustic guitar needs to radiate energy so that it can be heard. The bridge and top plate are designed to radiate efficiently. As the energy is radiated the string vibration is damped. The electric guitar does not need to radiate energy. The transducer that produces the electrical signal draws almost no energy from the string. The body of an electric guitar is solid and not designed to vibrate or radiate. Therefore, the string on an electric guitar is not damped by radiation.

25.6: It is true that the playing frequency depends on the timing of the stick–slip process, but the timing of the stick and slip depends on the motion of the kink along the string. The timing of the kink motion depends on the length of the string and the speed of sound on the string, not on the stickiness of the bow.

Chapter 26

26.1: $f_n/f_1 = [(2n + 1)/3.011]^2$. Therefore, $f_2/f_1 = 2.76$, $f_3/f_1 = 5.40$, and $f_4/f_1 = 8.93$.

26.2: Doing the squares leads to 81, 121, and 169. Forming the ratios leads to $81/81 = 1$, $121/81 = 1.494$, and $169/81 = 2.086$. The relative ratios are not changed by multiplying by 2 to get 2, 2.988, and 4.172. The listener interprets these spectral components as 2, 3, and 4 times a (nonexistent) fundamental frequency.

26.3: Adding a second drum head leads to a greater number of different frequencies in the radiation, essentially twice as many. The denser spectrum of unrelated components leads to a sound that is less tonal.

26.4: As your mother often told you, dinner plates of all kinds make bad musical instruments. Paper plates are particularly unsuccessful because of internal damping of vibrational waves in the plate material. A paper plate is soft, and vibrations in soft material are quickly converted into heat.

26.5: The modes are: [4,1], [4,1], [5,1], and [6,1].

26.6: The figure shows the standing waves on the rim of the bell for a mode with four meridian lines, like the "octave."

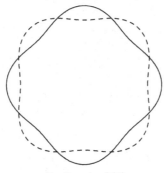

For Exercise 26.6

Chapter 27

27.1: A patch to create an amplitude-modulated signal is made from two sine function generators and a voltage controlled amplifier:

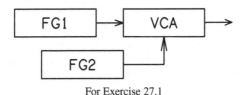

For Exercise 27.1

The control voltage to the voltage-controlled amplifier must be positive. Therefore, the output of function generator "2" must be offset so that it does not go negative.

27.2: $100,000,000/44,100 = 2267.6$. Hence, 2,267 operations, even without any parallel processing.

27.4: Pitch and pitch variation are essential elements in music expression. Guitarists bend notes and violinists play with vibrato. It's not surprising that synthesis techniques pay a lot of attention to pitch. (a) The depth of vibrato often increases as a tone is sustained. The modulation controller gives a performer the control to do that. (b) Without bending the pitch of notes, you cannot sing the blues ... or anything else.

27.5: A computer naturally handles data. A computer makes a great sequencer because programs can be written to display control data as musical notes and to edit the data. With a special interface, the computer can put out data at precise times. Computers do not excel in making waveforms. The sound cards inside personal computers are always a compromise.

♠

Index

W.M. Hartmann, *Principles of Musical Acoustics*, Undergraduate Lecture Notes
in Physics, DOI 10.1007/978-1-4614-6786-1,
© Springer Science+Business Media New York 2013